Metakaolin and Fly Ash as Mineral Admixtures for Concrete

Leonid Dvorkin

Professor and Head Department of Building Products Technology
and Material Science National University of Water and Environmental Engineering
Rivne, Ukraine

Vadim Zhitkovsky

National University of Water and Environmental Engineering
Rivne, Ukraine

Nataliya Lushnikova

National University of Water and Environmental Engineering
Rivne, Ukraine

Yuri Ribakov

Ariel University, Israel

T0144517

CRC Press
Taylor & Francis Group
Boca Raton   London   New York

CRC Press is an imprint of the
Taylor & Francis Group, an **informa** business

A SCIENCE PUBLISHERS BOOK

First edition published 2021
by CRC Press
6000 Broken Sound Parkway NW, Suite 300, Boca Raton, FL 33487-2742

and by CRC Press
4 Park Square, Milton Park, Abingdon, Oxon OX14 4RN

© 2021 Taylor & Francis Group, LLC
CRC Press is an imprint of Taylor & Francis Group, an Informa business

ISBN: 978-0-367-56214-4 (hbk)
ISBN: 978-0-367-56215-1 (pbk)
ISBN: 978-1-003-09682-5 (ebk)

Typeset in Times New Roman
by Radiant Productions

Preface

The use of mineral admixtures in cement concrete began in the second half of the nineteenth century with the beginning of its massive use in construction. Initially, based on experience of ancient Greeks and Romans volcanic rocks were used to obtain water—resistant mortars. These and other lime-active admixtures began to be used for increasing cement concrete water and corrosion resistance, especially in sea water, and then to modify their properties and save Portland cement. Active mineral admixtures have acquired particular importance in technology of new generation concretes, characterized by a number of improved construction and technical properties. Using such a highly active mineral admixture as microsilica in a composition with superplasticizers made it possible to radically improve concrete technological, strength and deformative properties as well as its durability. At the same time, production abilities for obtaining microsilica are limited, and there are difficulties with its transportation and dosing. Therefore it is logical to use it in modern concrete in a composition with admixtures of superplasticizers and other more easily available and cheap mineral admixtures, like fly ash and metakaolin.

The monograph presents results of experimental studies carried out by the authors to investigate the formation of the main properties of high-strength, reaction-powder, self-compacting and self-leveling concrete with complex admixtures, including fly ash, metakaolin and their compositions. Structure formation features and experimental statistical models are analyzed to quantify the influence of the main technological factors on the properties of concrete mixtures and hardened concrete. A methodology for the design of concrete compositions with organic mineral modifiers, including superplasticizers, fly ash and metakaolin is proposed.

The authors hope that the proposed technological solutions will be used in modern construction practice. The authors are grateful to colleagues who took part in experimental studies and in the preparation of this book for publication.

Contents

1
Introduction

In recent decades, new generation concretes have been increasingly used in construction. Their distinctive features are multicomponent, adjustable in a wide range of operational properties and have high manufacturability. Concrete of this type is characterized by a reduced specific consumption of cement per unit of strength when its high values are reached, high energy efficiency and the ability to minimize or eliminate vibration when compacting a concrete mixture and heat treatment in the manufacture of products and structures.

Active mineral admixtures are an important component of concretes. New generation concretes, in contrast to traditional ones, these admixtures are introduced into compositions with superplasticizers, which significantly increases their efficiency as a result of a significant improvement in the rheological properties of the cement matrix.

As numerous studies have shown, the greatest effect is achieved when a highly active mineral admixture-microsilica is introduced into concrete mixtures. At the same time, special production of microsilica as an admixture in concrete requires implementation of a rather complex energy-intensive technology of high-temperature processing of silica materials, sublimation and subsequent condensation of a finely dispersed amorphous material. Most commonly used is microsilica dust obtained during gas cleaning at enterprises for ferrosilicon and ferroalloys production. It has a limited distribution, its transportation, storage and dosing is fraught with certain difficulties.

This book contains research results on obtaining self-leveling, self-compacting, high-strength and reactive-powder concretes using such sufficiently available mineral admixtures such as fly ash, metakaolin and their compositions. In combination with superplasticizers, these mineral admixtures make it possible to obtain an effect commensurate with the effect of microsilica addition. The technological conclusions obtained by the authors can be used in the production of the above-discussed effective concrete types and the expansion of their use in construction.

Active mineral admixtures are an important component of concretes. In new generation concretes, in contrast to traditional ones, these admixtures are introduced into compositions with admixtures of superplasticizers, which significantly increases their efficiency as a result of a significant improvement in the rheological properties of the cement matrix.

2

Mineral Admixtures as Components of Cement Concrete

2.1 General Information about Mineral Admixtures for Concrete

Mineral admixtures are finely dispersed mineral materials, added into concrete mixtures in an amount that is usually higher than 5% to improve or provide special properties to concrete. By origin, admixtures of this type can be natural or technogenic. Mineral admixtures are divided according to their pozzolanic activity into inert and active. The group of active admixtures or pozzolan includes materials capable to react with calcium hydroxide at normal temperatures to form compounds with binding properties. Calcium hydroxide sources at concrete hardening are the main minerals that are part of Portland cement clinker and are hydrolyzed by exposure to water.

Self-cementing materials containing a significant amount of calcium oxide along with acid oxides (high calcium slags and ashes) are sometimes called active hydraulic admixtures, emphasizing their ability to slowly harden in an aquatic environment. For a more complete realization of such admixtures binding potential, additional lime content from an external source, such as Portland cement or lime, may be required. Classification of active mineral admixtures, used in European countries and USA, is based on their activity, chemical and mineralogical composition (Table 2.1).

Mineral admixtures are usually considered active if cement paste, made on the basis of the admixture and slaked lime, has a setting end no later than 7 days after mixing, and specimens from this paste remain water resistant 3 days after setting. Activity of admixtures is measured by the amount of CaO in mg, absorbed by 1 g of the admixture from a water lime mortar during 30 days. It ranges from 50 to 100 for fly ash and from 350 ... 400 mg of CaO or more per 1 g of highly active mineral admixtures (silica fume, metakaolin). The mineral admixtures activity is also determined using the microcalorimetric method according to the

Table 2.1. Classification and Characteristics of Mineral Admixtures.

No.	Classification		Chemical and mineral composition	Characteristic
	Feature	Material		
1	Binding properties	Quickly cooled metallurgical slag	Silicate glass containing CaO, MgO, Al_2O_3. Crystalline componentsin a small amount	Granules with 5 ...15% moisture. After drying, they are milled to a size <45 microns. S – 350...500 m²/kg (by Blaine)
2	Binding and pozzolanic properties	High calcium fly ash (CaO>10%)	Amorphous silica containing CaO, MgO, Al_2O_3. Crystalline components may be present as SiO_2, C_3A.	10...15% particles >45 μm. Most are spherical in shape. S = 400 m²/ kg (by Blaine)
3	High pozzolanic activity	Microsilica	Amorphous microsilica	Spherical shape particle powder d=0.1 μm. S ≈ 20 m²/ kg
		Rice husk ash	Same	Particles <45 μm with a developed cellular structure.. S ≈ 60 m²/ kg
4	Weak pozzolanic activity	Slowly chilled slag, hydraulic ash and ash slag	Crystalline silicate materials with a small amount of non-crystalline components	Additionally milled to provide pozzolanic properties
5	Normal pozzolanic activity	Low calcium fly ash (CaO <10%)	Silicate glass containing Al_2O_3, Fe_2O_3, alkalis Crystalline substance consisting of SiO_2, mullite, hematite, magnetite	Spherical particle powder> 45 μm. Most particles> 20 μm. S ≈ 250...350 m²/ kg (by Blaine)
		Natural materials	In addition to aluminosilicate glass, contains quartz, feldspar, mica	Most particles are grinded to a size of <45 μm and have an acute angle structure

specific heat and heat release at are wetting in polar and non-polar liquids, taking into account the hydrophilicity coefficient and other parameters.

Chemical methods for determining the mineral admixtures activity is insufficient to evaluate their quality. The most reliable method for comparative evaluation of mineral admixtures is to determine the Portland cement compressive strength under normal conditions or at accelerated hardening heat-moisture processing.

Fly ash is widely used as an active mineral admixture in concrete mixes. Fly ash is a finely dispersed product of the high-temperature processing of the mineral part of coal and other types of solid fuel. It is formed when they are burned in a dusty state in boilers furnaces and deposited by collectors. The most effective ash collectors are electrostatic precipitators that have efficiency of 95 ... 97%.

Due to the intensification of solid fuel combustion processes and using multi-ash types of coal and oil shale in thermal power engineering, the use of furnaces with liquid slag removal is effective. Liquid slag removal product from energy furnaces is fuel granulated slag formed as a result of rapid mineral melt cooling by water. Unlike ash, slag formed at higher temperatures practically contains no unburned fuel and is characterized by higher homogeneity. Slag is removed using hydraulic or dry methods. For a more common hydraulic method ash and slag are mixed.

Highly active mineral admixtures in concrete, which are increasingly used in recent decades, include ultrafine waste from ferroalloys production, the so-called microsilica (MS). Microsilica is a condensed aerosol collected by filters of gas purification systems of smelting metallurgical furnaces. It contains spherical shape particles with an average diameter of 0.1 μm (Fig. 2.1) and a specific surface area of 15 ... 25 m²/g and higher. Its bulk density is 150 ... 250 kg/m³. Chemical composition of MS is represented mainly by non-crystalline silica with a content that is usually 85 ... 98%.

Silica fume as an admixture in concrete was first proposed in the early 50s and began to be used in large quantities from the beginning of the 70s of the last century in Norway, and then in other countries. According to Norwegian standards, the amount of silicon dioxide in MS should be at least 85%, and the admixture content in concrete should be up to 10% by cement weight. The unique specific surface (up to 2000 m²/kg) in combination with an amorphized particle structure, the presence of impurities such as silicon carbide, which have high surface energy, determine the high structuring ability and reactivity of this material compared to other active mineral admixtures.

Due to its extremely high dispersion and amorphous particles structure microsilica causes a significant increase in the concrete mixtures water demand; therefore, it is used in combination with superplasticizers. It was showed that from the viewpoint of efficiency coefficient, taking into account the consumption of cement, superplasticizer and MS from various types of microsilica admixtures, waste of crystalline silicon and ferrosilicon production, containing SiO_2 in

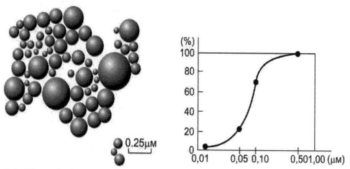

Figure 2.1. The main characteristics of microsilica: a—grains shape and size; b—granulometric content curve.

an amount above 89%, is most preferable. Wastes of silicomanganese and ferrochrome production are less effective.

Transportation of MS in an uncompressed dust state is quite difficult. Therefore it is usually granulated or used in a form divided into classes depending on the minimum allowable amount of SiO_2.

Based on MS, superplasticizer and some other admixtures, granular modifiers were proposed. These modifiers significantly simplify production of concrete with high technical properties, reduce the superplasticizer content, improve the concrete mixes consistency in time and provide a number of other advantages compared to separate the addition of admixtures. Along with MS, other mineral materials such as metakaolin, zeolites, etc., under certain conditions (high dispersion, combination with superplasticizers, etc.) can serve as effective modifiers of concrete.

In new generations of concrete, active mineral admixtures are used in combination with superplasticizers, achieving a radical increase in strength and an improvement in a number of other physical and mechanical properties. This book presents research results on the effectiveness of fly ash, metakaolin and their composition in combination with effective superplasticizers in self-compacting and high-strength and reactive powder concrete (RPC).

2.2 Interaction in the "Mineral Admixture (admixture)—Mineral Binder" System

The formation of cement composite building materials structure is a rather complicated process. It is a result of combined cement paste hydration and structure formation, i.e., a gradual transition from the coagulation structure to the formation of a spatial crystalline frame. To increase the strength of the filled cement concrete structure at the final stage of its formation, it is necessary to achieve the optimal concentration of the dispersed phase, and, if the particle size is optimal, ultimate packing and compaction of the system should be reached.

When mixing cement mixtures with water, dispersed admixture and cement particles are fixed in the spatial structural grid using coagulation contacts. For globular-type structures, the contact strength R depends on a number of factors:

$$R = Cf(F_p; \varphi; S^2_{sp}) \qquad (2.1)$$

where C is the chemical interaction constant;

F_p is resultant interaction force between particles;

φ is the filling degree by admixtures;

S_{sp} is the specific surface of particles, participating in the interaction.

The cement system strength is a result of synthesis of chemical processes, physico-chemical and physico-mechanical interaction, in which the admixture takes an active part.

Traditionally, three levels are distinguished in cement composite materials: micro, meso and macro levels. The microlevel is characterized by the molecular interaction of cement hydration products with admixture. An important feature of this process is the active participation of the admixture in chemical reactions. Chemically active admixtures shift the reaction direction towards intensive formation of hydration products, linking the latter into insoluble compounds. So, siliceous admixtures, interacting with calcium hydroxide Ca $(OH)_2$, form low-basic hydrosilicates. Calcium and magnesium carbonates interact with aluminum-containing clinker minerals, forming complex compounds of the type $3CaO \cdot Al_2O_3 \cdot (Ca, Mg) CO_3 \cdot 11H_2O$. Possibility of exchange reactions between carbonate admixtures and calcium hydrosilicates was also experimentally proven.

Achieving the maximum possible dispersion of the mineral admixtures usually has a positive effect on their properties. The unique chemical activity of microsilica dust fly-off, a waste from production ferrosilica is known. It is explained by an extremely high dispersion, more than 20 thousand cm^2/g. Having a high specific surface, mineral admixture, along with direct chemical interaction, affects the physicochemical processes at the interface. As the condensation-crystallization structure forms, epitaxial contacts form between the adhesive mass and admixture is formed. At the micro level, epitaxial contacts of close chemical nature grow during the direct completion of the admixture minerals crystal lattice.

According to the Gibbs–Folmer theory, the crystal nucleation energy significantly decreases in presence of crystallization centers, like admixture particles. The probability of spontaneous crystals nucleation b_κ can be expressed by the equation

$$b_k = e^{\frac{-\Delta W}{KT}} \tag{2.2}$$

where

$$\Delta W = -\frac{4}{3}\pi r^3 (Q\Delta T / T) + 4\pi t r^3 \sigma \tag{2.3}$$

Δ W is the change in free surface energy between solid and liquid phases; r is the radius of particles involved in crystals formation; K is the Boltzmann constant; T is the temperature; Q is the latent crystallization heat; Δ T is temperature difference at supercooling; σ is surface tension at the crystal-liquid interface and t is time.

Decreasing ΔW the mineral admixture substantially accelerates the crystallization of new products. Following Eq. (2.2), under heat-moisture processing conditions, filled cement systems exhibit a higher effect than for curing at normal conditions. Additionally, dispersing the admixtures decreases the grains radius and the surface tension at the crystal—liquid phase boundary due to electrolytes and other hardening accelerators, resulting in a much higher probability of a new phase creation. As the crystals of the new phase grow, the cement gel micropores are filled. As a result, at optimal admixture concentration and dispersion, a fine-grained binder structure is formed, which favorably affects the technical properties of artificial stone. It is established that at adding mineral admixture into the Portland cement-water system, the hardening rate and strength increase as long as all the admixtures grains remain surrounded by hydration products. For siliceous particles, the filling degree can be determined by the amount of CaO, which can bind 1 g of the admixture. The filling degree, calculated in this way, ranges from 5 to 10% of the cement weight.

Along with the constructive effect, mineral admixture can also have a destructive one. If the fillings with admixture parameters are beyond the optimum, in the entire volume of the cement stone or in its separate sections, which also possible, tensile stresses appear:

$$\Delta R_m = 0,5\alpha \Delta h R_m /(\varepsilon_m L) \tag{2.4}$$

where α is the hydration degree of the binder;

Δh is the thickness of a layer from several molecules wedged into the gap between the crystals and participating in the crystal completion;

R_m is the free crystal strength;

ε_m is the relative strain

L is the crystal size.

Thus, with an excess of admixture with a high dispersion of grains, local stress regions arise, which, as the crystals grow, can lead to cracking and other disturbances in the microstructure homogeniety. To avoid such destructive stresses, it is necessary to include in the admixture large-sized particles, which enables to form a new type of epitaxial contacts, based on electrostatic attraction.

The mesoscale is characterized primarily by physical interaction of the admixture particles, among themselves and with hydrated cement particles. Physical interaction of particles in hydration systems is substantially affected by cramped conditions, which differ by a sharp increase in the solid phase volume concentration and transfering part of the bulk water into a film one. Creation of cramped conditions allows for a technologically justified reduction of concrete mixture water demand. Using mineral admixtures at constancy of other parameters of the mixture contributes to creating such conditions at the coagulation stage of structure formation.

Having a size of one to three orders of magnitude higher than the lower limit of colloidal dispersion, admixture particles are fixed in the spatial grid by coagulation contacts forces. Ata sufficiently thin dispersion medium layer, they interact with each other.

The particle adhesion forces are:

$$F_c = \frac{2}{3}\pi Br / H^3 \tag{2.5}$$

where B is a molecular interaction constant;
H is the distance between particles.

However, their approach is prevented by the cement paste surface tension $F_{u.r.}$, proportional to the surface tension gradient and the binder (cement) particles radius r_c;

$$F_{c.p} = \pi r_c (2\sigma_{s.m} - \sigma_{s.v})$$ (2.6)

where $\sigma_{c.m}$, $\sigma_{c.v}$ are the surface tension at the solid boundary with a disperse medium and vacuum, respectively.

A feature of colloidal mortars is a relatively low kinetic and aggregative stability. Aggregate instability is associated with chemical transformations in a cement medium during transition to a condensation-crystallization structure and loss of plasticity. Kinetic stability is the property of particles to be held by surface forces in a suspended state without settling under the influence of gravity. Such stability is ensured if the following condition is satisfied:

$$mg \le \sum_{i=1}^{n} F_t$$ (2.7)

where m is the admixture grains weight; n is the number of contacts; F_t is the i-thtraction force.

In order for the particles to fall into the forces action field, it is necessary to fulfill the following condition $F_{ti} > F_{c.p}$ or to apply an external force F_{ext}, which together with the traction force F_c, would be greater than the ultimate viscosity of the paste. On the other hand, F_t is a parameter that depends mainly on the distance H. The average distance between the admixture particles H_{av} is determined by the ratio of the average matrix volume V_m to the total particle surface area:

$$H_{av} = V_m /(V_f S_{sp.})$$ (2.8)

where V_f is the volume and S_{sp} is the admixture specific surface.

Without external influence maximum contact strength is not provided. Such external influence is intensive compacting filled with mineral admixture systems. However, it is energy and time consuming. Therefore, changing the cement paste viscosity parameters due by using plasticizing SAS is more promising. Adsorbed on the particles surface, SAS shields them, changing or eliminating the electrostatic interaction. This process causes the admixture particles aggregation during the binder transfer into a thin-film state.

The macro level is characterized by cement concrete formation on two sublevels: (1) filled with mineral admixture cement paste—aggregate; (2) admixture—cement paste. At macro level the mineral admixture should provide maximum adhesive strength between binder and aggregate, maximum binder cohesive strength, and minimal porosity due to the crowding out of the cement paste into the contact zone. Moreover, the filling degree should be such that at the initial stage of the structure formation, the specified rheological parameters of the mixture are provided.

Strength of the contact zone $R_{c.z}$ at other equal conditions is the main structural indicator that determines the concrete strength. The main parameter characterizing $R_{c.z}$ is the thickness of the cement stone layer $\delta_{c.s}$. Formation of this layer under the influence of surface energy fields, chemical and mineralogical composition of the aggregate is different, compared to that in a volume. The optimal thickness $\delta_{c.s}$ is obtained from the condition of ensuring the maximum volumetric concentration of the aggregate particles framework that satisfies the rheological and strength properties requirements. The distance between adjacent elements of the rigid frame is governed mainly by the specific content of cement in concrete and the water-cement ratio.

2.3 Activation of Mineral Admixtures

Analysis of contact interactions mechanisms in filled with mineral admixtures cement systems makes it possible to outline ways for activating admixtures in order to enhance their adhesion to the binder and increase the structure-forming role. Creation of sufficiently strong adhesive contacts in the cement-admixture system is possible only if the admixture surface energy is much higher than that of cement. This conclusion is based on the thermodynamic concept of adhesion, according to which the main role in adhesive strength formation is assigned to the ratio of the adhesive surface energy W_{ad} and the substrate W_{sub}. In this case, the obligatory condition is

$$W_{ad} < W_{sub} \tag{2.9}$$

Determining the surface energy of solids is difficult. For liquids, the equivalent of the surface energy concept is surface tension, which is determined by well-developed experimental methods. Its value for solids can be evaluated indirectly using a series of calculation methods or measuring some mechanical characteristics. For example, to assess the minerals surface energy, methods of grinding, drilling, scratching, damped vibrations are widespread. Applying methods also based on determining the fracture energy, are based on Griffiths theory.

To assess the change in surface energy upon powdered admixtures activation, of interest are methods based on wetting in the "solid-liquid" system, in particular, the method for determining the critical surface tension σ_c, requiring experimental determination of a solid body wetting angle cosine θ, depending on the surface tension of the liquid. The range of surface energy values for different materials is very wide: from 0.072 J/m^2 for water at normal temperature to $1 \ldots 2 \text{ J/m}^2$ for such materials as diamond or silicon carbide.

The ways of mineral admixtures physicochemical activation to a large extent follow from the Dupre - Young equation, which additionally takes into account the vapor adsorption effect and the solid surface structuring effect,

$$W_{ad} = W_s - W_s^* (m + \cos \theta) \tag{2.10}$$

where W_{ad} is the adhesion work;

W_s –is surface energy of solid;

W_s^* is free surface energy of a solid in vapor or gas atmosphere;

$m = \sigma_l/\sigma'_l > 1$ (the ratio of the surface tension of the liquid to the value of this parameter oriented under the action of the force field of the solid surface), θ - wetting angle.

Surface energy is a component of the total energy of a solid, which is determined by the total effect of atomic vibrations energy, kinetic energy of chaotic translational and rotational motion of microparticles (molecules, atoms, ions, free electrons, etc.), potential energy of interaction of these particles, energy of atoms and ions electronic shells, intranuclear energy, electromagnetic radiation energy.

According to the Gibbs—Helmholtz equation, the free specific surface energy is obtained as follows:

$$U = \sigma - T\partial\sigma/\partial T \tag{2.11}$$

where U is total surface energy;

T∂σ/∂T is heat of surface unit formation.

For crystalline solids, the specific surface energy depends on the crystal cell strength, as well as on properties of the medium surrounding the body. An effective way to activate admixtures by increasing surface energy is mechanochemical processing. An increase in surface energy is caused, first of all, by breaking of interatomic bonds of the structure, occuring at solids crushing, grinding or abrasion. Newly formed surfaces have a significantly higher surface energy value, which leads to their higher adhesive activity.

Mechanical processes during the grinding of mineral and organic materials causes, along with an increase in their surface energy, an increase in the powders isobaric potential and, accordingly, their chemical activity, which also contributes to high adhesive strength at contact with binders. However, the tendency of milled powders to rapidly deactivate in air as a result of high adsorption ability and mutual compensation of the formed charges should be taken into account. The lifetime of the radicals arising during mechanochemical processing in the air is only $10^{-3}...10^{-6}$—s. Adsorption of the freshly milled powders of moisture and carbon dioxide vapors from the air and saturation of uncompensated molecular forces leads to "aging" of the admixtures surface and also serves as an additional obstacle to the formation of reliable adhesive contacts. Therefore, mechanochemical activation of mineral admixtures is effective when creating a primary contact layer of a structured binder on their grains directly during grinding.

Activation of admixtures adhesion ability by increasing their free surface energy can be achieved by exposure to electric and magnetic fields, ultrasonic treatment, using ionizing radiation.

It is obvious from Eq. (2.9) that in order to achieve high adhesive strength it is important to provide the necessary wettability of the admixture by binder. Wetting is thermodynamically possible if

$$\sigma_s > \sigma_{s.l} \tag{2.12}$$

where σ_s is the solid body surface energy;

$\sigma_{s,l}$ is interfacial surface energy at the solid – liquid interface.

An effective way to reduce interfacial surface energy is to treat admixtures by SAS.

The decrease in interfacial surface energy when creating an adsorption-active medium is determined by the following equation

$$\Delta W_{s,l} = KT \int_0^c n_s(c) d \ln c \qquad (2.13)$$

where $\Delta W_{s,l}$ is the difference in interfacial surface energy without surfactants and in the presence of surfactants with concentration C;

n_s is adsorption, determined by the number of surfactant molecules adsorbed per 1 cm^2 of the interface;

K is Boltzmann constant;

T is absolute temperature, $°K$.

One of the main SAS parameters can be the ratio λ characterizing the hydrophilic-lyophilic balance:

$$\lambda = (b + \varphi_1 n_{CH_2})/a' \qquad (2.14)$$

where $b + \varphi_1 n_{CH_2}$ is the SAS non-polar molecule or ion affinity to a hydrocarbon medium (n_{CH_2} is number of CH$_2$ links); a' is the polar group affinity to water.

When choosing SAS, the chemical nature of admixture and binder must be taken into account. In particular, if admixtures are salts of alkaline earth metals (carbonates, silicates and aluminosilicates of calcium and magnesium) for cement and other systems with a hydrophilic binder, it is advisable to use anionic SAS.

A necessary condition for SAS effectiveness is their ability to chemisorptions interact ions with the admixture particles surface. In a general case, for mineral admixtures of an acidic nature, SAS of the cationic type are most effective, and the basic nature admixtures anionic are SAS are more suitable.

The influence of the adsorption-active medium on the ΔW value becomes higher with admixture dispersion and concentration growth, which is associated with an increase in the interfacial surface and, correspondingly, with surface energy excess. The minimum surface energy value at the interface is achieved if the binder and admixture molecular natures are close. The value of $\sigma_{s,l}$ decreases as the difference in solid and a liquid polarity becomes smaller.

Along with activating the mineral admixtures adhesive interaction with binder, methods of activating admixtures to intensify of the binder epitaxial crystallization are of undoubted interest. For example, with respect to cement systems, admixtures are more preferred substrates for hydrate nuclei formation than the initial cement particles. Two-dimensional hydrate nuclei are firmly fixed on the admixture surface and intensify the organized growth of the cement stone structure in the direction perpendicular to their particles surface. The nuclei of neoplasms crystals formed on the initial cement particles, as a result of their dissolution, are washed off by water and are suspended in it, which contributes to organization of a random, disorganized cement stone structure.

The mineral admixtures effectiveness as substrates of directional crystal formation increases with dispersion growth, at crystallochemical proximity to the binder and at adding activating chemical admixtures. As such activators, in particular, that are mostly used are salts-sulfates, phosphates, calcium fluorides and substances containing high oxidation degree elements.

2.4 General Characteristics of Fly Ash as Active Mineral Admixture of Cement Concrete

Fly ash (here in after referred to as ash) is a finely dispersed product of the mineral part of coals at high-temperature processing. It is formed when these are burned in a dusty state in the furnaces of boilers and then deposited by trapping devices from chimneys. The most effective ash collectors are electrostatic filters that have an efficiency 95...97%. Ash captured and transported in a dry form is preferable as an active mineral admixture.

The main component of ash (65%) is the glassy aluminosilicate phase in the form of spherical particles that are up to 100 µm in size. Of the crystalline phases ash can include α-quartz and mullite, and in cases of high Fe_2O_3 content—also hematite. The quantitative ratio between α-quartz and mullite is determined by that of SiO_2/Al_2O_3, with an increase in which the content of α-quartz in crystalline form increases, and the content of mullite decreases. Correspondingly, the activity of the ashes by lime absorption increases. Ashes, enriched by iron oxide, are more fusible, they form more glass on the particles surface and lime absorption activity increases. The ash activity depends on the glass phase content. There is a close relationship between ash containing mortar strength and the glass phase specific surface. Glass in ashes can be considered to be a material that contains amorphites—formations that are closely related by composition and structure to the corresponding crystalline phases, but with a very high specific surface, and disordered alumina-silica interlayers between them. The ability of the glass phase to hydration and hydrolyze can be explained by a loose submicrostructure and relatively high permeability of amorphites, due to the voids between ionic groups. The activity of the glass phase intermediate substance is determined by the ratio of alumina and silica: the higher it is, the easier the fly ash hydration process in alkaline and sulfate-alkaline environment, in a neutral environment it is stable. Hydraulic activity of calcium-aluminosilicate glass contained in the ash is positively influenced by magnesium, iron and some other elements.

A certain hydraulic activity in ash, along with the glass phase, has dehydrated and amorphized clay substance. Activity depends on the mineralogical composition of clays that are mineral part of the fuel. With an increase in the amorphized clay substance content in ash, its water demand increases. If the mineral part of the fuel has a significant carbonate content, low-basic silicates, aluminates and calcium ferrites, able to interact with water, are formed in ash.

Brown and hard coals of a number of deposits in Central Asia and Siberia, as well as oil shale have a lesser amount of high-calcium ash. Ash also contains a small amount of free calcium and magnesium oxides, sulfates, sulfides, etc.

Ashes, as a rule, contain carbon in a form of various coke residues modifications, the content of which depends on the type of burned fuel: it is less than 4%, for brown coal and oil shale, 3 ... 12%, for hard coal and 15 ... 25% for anthracite. The content of unburned carbon particles (UCP) in the finely dispersed fractions of ash is less than in coarse particles. Chemical composition of ashes varies depending on coal deposits. The content of basic oxides in the ashes of various thermal power plants is within the following limits: $SiO_2 - 37 ... 63$ %, $Al_2O_3 - 9 ... 37$ %, $Fe_2O_3 - 4 ... 17$ %, $CaO - 1 ... 32$ %, $MgO - 0.1 ... 5$ %, $SO_3 - 0.05 ... 2.5$ %, $Na_2O + K_2O - 0.5 ... 5$ %. Losses at ignition, characterizing the content of UCP in ash, are 0.5 ... 3%. Chemical components with highest content are SiO_2 and Al_2O_3 are presented predominantly in the glass phase, the main part of SiO_2 is in a form of quartz, and Al_2O_3 is in form of mullite ($3 Al_2O_3 \cdot 2 SiO_2$). The chemical composition of ash varies significantly even when burning the same fuel at the same power plant. The average chemical composition of ash at each thermal power plant is usually quite stable.

Important indicators of ash quality are the dispersion and particle size distribution. Numerous studies show that there is no direct correlation between these two indicators. Ash dispersion is expressed by specific surfaces, determined by air penetration method, as well as by residue on sieves at sieving. This indicator varies from 1000 to 4000 cm^2/g, in many cases it approaches the specific surface of cement. Ashes containing more residues of non-combustible fuel have higher specific surface values.

Granulometric composition of ashes is different. The grain size is within the range of 1 ... 200 microns. The content of fractions > 85 microns is up to 20%, that of 30 ... 40 microns fraction is about 50%. More coarse ash fractions are formed when the content of CaO and Fe_2O_3 oxides of fluxes in the mineral part of fuel is increased. The ash dispersion depends on the fineness of pulverized fuel grinding, with a decrease in the latter, the amount of unburned particles increases. The most dispersed ash is captured by electrostatic filters, while the ash granulometric composition varies for different electrostatic filters fields.

Different ash fractions have different true and average densities. This is explained by differences in chemical and mineralogical composition and particle shape. The ash density decreases with increasing the grains size. For example, ash of the Dnieper state district power station (Ukraine) contains 27% of carbon particles > 85 microns, and about 9% of fractions smaller than 60 microns. The average bulk density of ash ranges from 600 to 1100, the true density is from 1800 to 2400 kg/m^3.

Ash is characterized by a significant content of particles having small closed pores. These pores are formed due to swelling of molten mineral mass with gases released during the clay minerals dehydration, dissociation of limestone, gypsum and organic substances particles. Pores can reach 60% of the ash particles volume. High content of micropores in ash yields a high specific surface value, which is an order of magnitude higher than that of cement. High specific surface area of ash is associated with such properties as adsorption capacity, hygroscopicity, and hydraulic activity.

Following the classification of ashes and slags, based on genesis principle, ashes belong to the group of materials obtained by solid-phase reactions and the interaction of solid phases with melt. They are divided into high calcium ($CaO > 20\%$) and low calcium ones ($CaO < 20\%$). Crystalline phases predominate in the first, whereas glass and amorphized clay material—in the second. High-calcium ash is divided into low-sulfate ($SO_3 < 5\%$), obtained by burning coal and peat, and sulfate ($SO_3 > 5\%$)—by burning shale.

Ashes, like other fuel-containing waste, can be classified:

- By the type of fuel as those received at processing and burning brown and bituminous coal, anthracites, oil shale and peat;
- by their plasticity as high-, medium-, moderately plastic and unplastic;
- by the combustible part content as small (up to 5%), medium (5 ... 20), high (more than 20%);
- by their chemical composition—with a high content of acidic oxides and alkalis, oxides of calcium, magnesium and sulfur;
- by their melting degree—high-melting (melting point > 1400°C), medium-melting (1250 ... 1400°C) and low-melting (< 1250°C);
- by softening interval (i.e., difference between the melting temperatures and specimens deformation initiation)—with short and long softening intervals (< 50°C and > 50°C, respectively);
- by their swelling degree—weakly or non-swelling, medium and highly swelling.

An integral characteristic of ashes chemical composition is basicity modulus (m), which is the weight ratio of basic oxides to acidic fractions. For basic ashes $m > 0.9$, for acid ones $m = 0.6 ... 0.9$, for super acid ashes $m < 0.6$. In basic ashes, the total CaO content reaches 50, and in super acids—12%. The latter are the most common.

Ashes are classified by unburned carbon content (UCC), %, into 6 categories:

1	2	3	4	5	6
Up to 5	6 ... 10	11 ... 15	16 ... 20	21 ... 25	> 25

By specific surface area, ash is divided into fine ($S_a > 4000$ cm^2/g), medium ($S_a = 2000 ... 4000$ cm^2/g) and coarse ($S_a < 2000$ cm^2/g). If bulk density is < 800 kg/m^3, ash is considered light; for 800 ... 1000—medium density and for > 1000 kg/m^3—heavy. To characterize ashes as active mineral additives, their hydraulic activity is important. Traditional methods to determine the hydraulic activity of ash is by lime absorption from lime mortar and the ability of ash to exhibit binding properties when interacting with hydrated lime. Materials hydraulic activity can be also determined using the micro-calorimetric method by the value of their wetting heat in polar and non-polar liquids, taking into account the hydrophilicity coefficient and a number of other parameters.

Requirements with ash as the active mineral additive are determined by physicochemical mechanism of their influence on hardening of concrete and its structure formation. Hydraulic activity of ash, as well as other pozzolanic type substances, is largely due to the chemical interaction of silicon and aluminum oxides, contained in them, with calcium hydroxide, released during hydrolysis of clinker minerals, with formation of calcium hydrosilicates and hydroaluminates. Hydraulic activity of ashes is significantly affected by the latent heat of their crystallization activity, which is manifested by a differential thermal analysis. Ashes hydration is facilitated by their glass phase; as the crystalline phase is practically inert. Chemical activity of ashes is directly related to their dispersion.

The strength of cement and concrete with the addition of ash depends on thickness of the ash particle surface layer, affected by chemical processes. Various factors, including ash dispersion, and glass content, have an effect as they accelerate the surface corrosion of the ash particles placed in the cement stone. Some researchers attribute the positive effect of ash on the concrete structure formation to the "fine powder effect", expanding the free space in which hydration products precipitate, accelerating cement hardening process.

The type of burned coal ash can be divided into:

- anthracite, produced by burning anthracite, semi-anthracite, lean coal;
- coal, obtained by burning coal (except lean) coal;
- brown coal fly ash formed during the burning of brown coal.

Depending on the application field, ash is divided into the following types: (1) for reinforced concrete structures and elements; (2) for concrete structures and elements; (3) for hydraulic structures, as well by classes—for heavy (A) and light weight (B) concrete.

Specific surfaces of class A ash should be at least 2800 cm^2/g, and class B ash should be in the range of 1500 ... 4000 cm^2/g. The residue on a No. 008 sieve for Class A ash should not exceed 15% by weight.

By chemical composition and humidity, ash must satisfy the requirements specified in Table 2.2. According to modern design provisions, ash should be provided a uniform change in volume within its mixture with cement when testing samples by boiling in water.

If the uniformity in the samples size change when testing them in an autoclave at a pressure of $(2.1 + 0.1)$ MPa is provided, the content of free calcium and magnesium oxides in ash in amounts exceeding those specified in Table 3.1 is allowed. Samples are considered to have passed the test if their elongation does not exceed: 0.5% and 0.2%.for concrete on Portland cement and slag Portland cement, respectively.

Most standards for ash also limit its moisture, chemical composition and dispersion. The main indicators set in this case vary within certain limits: humidity less than 1 ... 3%, loss on ignition—up to 5 ... 12%, SiO_2 content—at least 40 ... 45%, SO_3—up to 2.5 ... 5%, dispersion—at least 1250 ... 4250 cm^2/g (Table 2.2). In all cases when ash is used in concrete mix prepared using reactive aggregates, alkali content should be considered. Presence of UCC in ash causes an increase in concrete mix water demand. The negative effect of carbon particles on concrete strength decreases as the curing duration increases. It is found that when using ash with a carbon content of 2.9; 13.85 and 15.55%, the concrete strength remains practically constant after 18 months of hardening.

It is rational to associate the possibility of using ash in concrete production with UCC concentration in the ash-cement stone rather with their content in ash. In this case, it is possible to use ash with a UCC of up to 20 ... 25%.

2.5 Concrete Proportioning with Fly Ash Admixture

The selection of concrete compositions with the addition of ash should consist in determining a ratio of components in which the required properties of the concrete mixture and hardened concrete are achieved at minimum cement consumption.

Ash plays in the concrete mix the role of an active mineral additive that increases the total amount of binder and also a microfiller, which improves the granulometry of sand and actively affects the structure formation processes of concrete. Given the ash additive's polyfunctional nature, the replacing by ash a part of cement or sand does not allow us to solve the compositions optimization problem.

A decrease in cement consumption by using ash is advisable at excessive cement activity, i.e., in cases where the cement brand is higher than recommended. When using ash, it is allowed to reduce the minimum standard cement consumption rate for unreinforced concrete products to 150 kg/m, and for reinforced concrete—to 180 kg/m, the total consumption of cement and ash should be at least 200 and 220 kg/m^3, respectively.

To determine the amount of ash at excessive cement activity, a calculation-experimental technique was proposed [1]. If the normal densities of ash and cement pastes are close and concrete hardening is at normal conditions, it is assumed that the ash content should be assigned proportionally to the percentage of excessive cement activity reduction.

If the water demand of ash exceeds 30%, the mixture content should be decreased by the following coefficient

$$K = Q_{c.p.}/Q_a \qquad (2.15)$$

where $Q_{c.p.}$ is the normal consistency paste output per 1 g of cement, cm^3,

Q_a - same per 1 ton of ash.

The paste output $Q_{c.p}$ is obtained as

Table 2.2. Requirements to Ash as Concrete Admixture.

Feature	The indicator value for ash type (class)		
	I (A and B)	II (A and B)	III (A)
Content of $SiO_2+Al_2O_3+Fe_2O_3$, % by weight, at least, for ash:			
anthracite, coal	70	Not normed	70
brown coal	50	Same	50
Content of sulfur and sulfate compounds in terms of SO_3,% by weight, not more than	3	3.5	3
Content of free calcium oxide (CaO),% by weight, not more than	3	5	2
The content of magnesium oxide (MgO),% by mass, not more than	5	5	5
Loss on ignition,% by weight, not more than, for ash:			
anthracite	15	20	5
coal	7	10	5
brown coal	5	5	3
Humidity,% by mass, no more	3	3	3

$$Q_{c.p} = 1/\rho + N.C./100 \qquad (2.16)$$

where ρ is the true densities of cement or ash, g/cm^3,

N.C. is the normal paste consistency.

This technique is recommended for selecting four concrete compositions, allowing determining the final ash content, considering cases of 5% less and more than the calculated. Ash content, at which the required concrete properties are provided at lowest cement consumption, is considered as optimal. If the lowest cement consumption is at maximum ash content, then one or two more concrete compositions with ash contents increased by 5 ... 10% are selected. The main rule in this technique—the proportionality of ash consumption to the necessary decrease in cement activity—is theoretically poorly substantiated, which leads to significant deviations of the calculated values relative to the experimental ones.

An alternative technique for concrete compositions design is based on considering the cement consumption efficiency coefficient K_e. When using ash

$$K_e = f_c/C \qquad (2.17)$$

where f_c is concrete strength at given age, MPa,

C is cement consumption, kg/m^3.

The value of K_e is obtained empirically. For steamed concrete without ash at a water-binding ratio in the range of 0.42 ... 0.54 after one day of hardening K_e ranges from 0.038 to 0.058. If the ash content is 60%, K_e varies from 0.067 to 0.12. For concrete after 28 days of normal hardening K_e increases accordingly from 0.065...0.085 to 0.067...0.12.

To obtain concrete with equal strength at selected mixed binder composition, cement consumption can be determined as follows:

$$C = f_c/K_e \qquad (2.18)$$

Ash content in this case is

$$A = C \, m_a/100 - m_a \qquad (2.19)$$

where m_a is the part of ash in the mixed binder.

Currently there is not enough experimental data to confirm this method.

Table 2.3 shows our data on the values of efficiency coefficient for the ash of Ladyzhin state district power station (Ukraine) in concrete with equal workability (cone slump CS = 4...5 cm), produced using Portland cement (class 32.5). The cement contained up to 20% of blast furnace granulated slag, its normal consistency was 24 ... 26%, and the average strength at steaming was 25 MPa. Quartz sand with a particle size of 1.4 ... 1.5 mm was used as the fine aggregate. The coarse aggregate was used, crushed granite fractions of 5 ... 20 and 20 ... 40 mm. The chemical composition and basic physical properties of the ash are given in Table 2.4.

As it follows from Table 2.3, for all of the investigated concrete compositions, a significant increase in K_e is observed as the ash content increases to 150 kg/m³. The most intensive growth of K_e is obtained for steamed concrete immediately after thermal processing. A relatively lower increase in K_e is obtained for concretes at normal hardening and as the concrete class increases. Regardless of the concrete class, the maximum value of K_e is evident at ash content of 0.1 ... 0.12. For concrete without ash the value of the coefficient for classes C8/10...C16/20 is approximately 0.03, while for concrete class C20/25 and higher it is more than 0.07.

Table 2.5 presents the recommended nominal concrete compositions with addition of ash, allowing the achievement of the required concrete class both at normal hardening and at heat processing. In the latter case, 70% of concrete strength is achieved after steaming. Analysis of these compositions shows that adding ash allows for saving up to 25% of cement at equal concrete strength. It should be noted that if the ash content in concrete is at least 150 kg/m³, the recommended lower limit of cement consumption, can be reduced.

Approximate compositions of heavy concrete includes 150 kg/m³ of ash, crushed granite stone fraction 10 ... 20 mm and fine-grained quartz sand. Such concrete mixtures with cone slump of 2... 5 cm under heat treatment conditions require 115 ... 130 kg/m³ of Portland cement. Cement consumption for concrete mixtures with stiffness of 30 ... 60 s (by technical viscometer) is 90 ... 100 kg/m³.

Additional strength due to the addition of ash was obtained by the authors for concrete class C8/10 at Portland cement CEM II 32.5 consumption of 150 kg/m³ (Table 2.6).

Addition of ash into concrete mix, unlike other active mineral additives, usually improves its workability. The first two authors of this book have already demonstrated that dependence of concrete mixture workability on ash content has an extreme form, and optimal ash content should be less than 30% of the binder weight [2]. The plasticizing effect of ash is influenced by the particles surface state and their dispersion. Concrete mixture workability is improved when ash is added due to the glass surface of its constituent particles, reducing internal friction in the concrete mixture and its viscosity.

A number of researchers [2, 3] believe that spherical ash particles can be considered as solid "ball bearings" in a mixture, similar to emulsified air bubbles, when using air-entraining admixtures, which have a plasticizing effect on concrete mixture. Coarse fractions contain more UCC with increased water absorption and irregularly shaped particles, therefore, when using ashes of a high dispersion the water demand is significantly reduced.

Increasing the ashes dispersion and reducing their water demand can be achieved by selecting them from the last units of electrostatic filters or grinding, destroying their organomineral aggregates. Decrease in ash water demand at grinding is explained by decreasing the amount of capillary water held by aggregated particles, which leads to a more significant effect than the increase in the amount of adsorbed film water occurring with an increase in the specific surface.

Addition of ash reduces the concrete mixture water segregation [2]. The plasticizing and water-holding ability of ash determines the promise of its use in lightweight concrete. Concrete mixtures with optimal ash content have a rather high stability and are suitable for transportation over long distances. At the same time, it is known that a high ash content helps to accelerate the setting time.

The influence of ash on concrete strength depends on its properties and dispersion, content, chemical and mineralogical composition of cement, concrete age and processing conditions. To assess the effect of ash on concrete strength, its "cementing efficiency" concept, characterized by coefficient $K_{c.e}$, is introduced. When predicting the concrete strength, it is proposed to find the modified cement-water ratio as [4].

$$(C/W)_m = (C + K_{c.e.} m_a)/W \qquad (3.6)$$

where C is the cement consumption in concrete with ash addition;

W is the water demand;

m_a is the weight part of ash in mixed binder.

Neither the carbon content nor specific surface can be used to estimate the $K_{c.e.}$ of ash. Ashes, having the lowest [5] specific surface area and various carbon contents, showed the highest values of $K_{c.e.}$ coefficient. In order to determine the $K_{c.e.}$ value it is possible to use experimental dependence $f_c=f$ (C/W) for a given cement and then, according to the strength values of concrete with addition of ash, determine $(C/W)_m$. It was found that for ash of the Ladyzhin state district power station (Ukraine) the value of $K_{c.e.}$ coefficient ranges from 0.2 to 0.4 [6].

Having determined the value of $(C/W)_m$ and optimum ash content with the known $K_{c.e.}$ the required C/W of ash-containing concrete can be found and

Table 2.3. The effect of Ash Content on Concrete Strength.

Content, kg/m³		W/(C+A)	Compressive strength, MPa			Efficiency coefficient		
Cement	Ash		After steaming, f_c	28 days after steaming f_c''	28 days of normal hardening f_c'''	f_c/C	f_c''/C	f_c'''/C
150	-	1.12	4.4	6.4	7.8	0.029	0.043	0.052
150	50	0.87	9.7	11.6	11.4	0.065	0.077	0.076
150	75	0.78	11.6	14.0	10.6	0.077	0.093	0.072
150	100	0.70	14.2	14.9	12.5	0.095	0.099	0.083
150	125	0.63	14.4	16.1	15.6	0.096	0.107	0.104
150	150	0.58	15.7	16.6	14.7	0.105	0.110	0.098
150	175	0.54	15.6	16.8	15.3	0.104	0.112	0.102
150	200	0.50	16,1	17,3	16.6	0,107	0,115	0,111
250	-	0.72	8.5	12.5	13.3	0.034	0.050	0.053
250	50	0.60	11.5	15.4	13.9	0.046	0.062	0.056
250	100	0.52	24.0	26.6	24.2	0.096	0.106	0.097
250	150	0.45	26.1	29.1	25.4	0.104	0.116	0.102
250	200	0.40	27.0	29.6	26.6	0.108	0.118	0.106
370	-	0.45	29.1	37.7	34.5	0.079	0.102	0.098
370	75	0.40	37.9	46.0	47.1	0.102	0.124	0.127
370	100	0.38	38.2	43.7	44.6	0.103	0.118	0.120
370	125	0.37	33.9	38.0	32.3	0.092	0.103	0.087
370	150	0.35	40.8	43.7	39.5	0.110	0.118	0.106
370	175	0.35	35.0	39.2	33.4	0.095	0.106	0.090
370	200	0.33	37.8	37.6	34.2	0.102	0.101	0.092

Table 2.4. Composition and Properties of Ash used for Experiments, %.

Component	Following the data from Power stations		
	1	2	3
SiO_2	56.3	54.43	55.3
TiO_2	1.1		1.4
Al_2O_3	25.1	26.6	22.34
Fe_2O_3	9,2	8,72	5.42
FeO	-	-	2.52
MgO	1.4	1.61	0.12
MnO	-	-	2.46
CaO	2.3	1.31	5.96
Na_2O	1.4	0.64	0.75
K_2O	2.7	3.84	2.46
SO_3	0.5	0.31	0.38
P_2O_5	-	-	0.33
Loss on ignition,%	1.6	0.53	0.24
Properties			
True density, g/cm^3	-	2.32	-
Specific surface, cm^2/g	3432	3303	-
Sieve residue №008,%	-	18	-
Water demand at normal paste consistency,%	-	15	-

corresponding compositions can be designed. To achieve the same workability, it is required [7] that the ratio between the free water volume, used above the amount necessary to achieve the ash and cement powders normal consistency, to that of the entire binder in the compared concrete mixtures should be the same amount. Additionally, in mixtures of the same workability, the volume ratio of paste, based on a cement-ash binder and aggregate, should be the same.

The most favorable effect of ash additive on concrete strength is at relatively small binder consumption. This is due to a noticeable decrease in water demand for low-cement concrete mixes when replacing part of cement by ash [8].

Most researchers proved the positive effect of increasing ash dispersion on concrete strength. It has been shown in [9, 10] that ash activity increases significantly when its particles size is increased to 5 ... 30 μm [9]. The product of an ash specific surface and its glass phase content is close to coefficient K in the Fere formula. This coefficient is directly proportional to concrete strength. According to the Fere formula, concrete compressive strength at 28 days is

$$f^{28}_c = K(V_c/(V_c + V_w + V_{air}))^2 \qquad (3.7)$$

where V_c, V_w and V_{air} are volumes of cement; water and air, respectively.

Table 2.5. Recommended Compositions of Normal Weight Concrete with Ash.

Concrete strength, MPa	Cone slump, cm	Cement strength, MPa		Materials consumption per 1 m³, kg					Water demand, l
		40	50	Ash	Sand	Crushed stone fraction, mm			
						3...10	5...20	20..40	
10	1...3	200	-	-	710	-	540	800	150
		150	-	150	570	-	400	970	160
15	3...5	240	-	-	700	-	500	800	165
		200	-	150	570	-	400	880	180
20	3...5	290	260	-	600	-	560	800	165
		225	200	150	490	-	400	960	180
20	7...9	360	330	-	620	-	1280	-	170
	6...8	280	250	150	460	-	1330	-	180
30	4...6	350	290	-	530	-	510	800	170
	5...7	340	310	150	400	-	400	940	180
30	7...9	420	380	-	540	-	1250	-	190
	5...7	370	370	150	400	-	1300	-	180
30	1...3	330	300	-	530	-	1280	-	180
	1...3	300	270	150	400	-	400	970	175

Having examined the strength of cement mortars, obtained by mixing clinker and ash that were miled up to specific surface of 2500...6400 and 3000...8000 cm²/g, respectively, M. Venuat has found the necessary correspondence between the ash granulometric composition and clinker grinding fineness of [11]. The most significant effect of increasing ash dispersion on concrete strength is obtained at an early age.

Compared to separate grinding, the best results were obtained by the co-grinding of cement and ash. Concrete binder with 30% ash with a specific surface of 5000 cm²/g. had at 28 days, a higher compressive and tensile strength than that of Portland cement without additives [6]. Even when 40% of cement was replaced by ash, concrete strength at 28 days was close to that without additives, and after 60 days almost no difference in strength values was observed.

A significant effect of increasing dispersion is observed also after heat-moisture treatment of concrete. This effect is weakened at 28 days [3].

The effect of ash dispersion on concrete strength is noticeably stronger compared to that of cement. This is due to the noticeable plasticizing effect of fine ash fractions on concrete mixtures, despite an increase in the normal consistence of ash-containing cements [9].

To achieve the high strength of ash-containing concrete, the chemical and mineralogical composition of the clinker is of particular importance. At early age an increase in concrete strength is caused by increased alkalis content in clinker, accelerating chemical interaction between ash and cement. For exhibiting pozzolanic reaction of ash at a later age, cements with a high alite content are preferred, as these cements form a significant amount of $Ca(OH)_2$

Table 2.6. Increase in Concrete Strength with Addition of Ash.

| Components content, kg/m³ | | | | | | Compressive strength, MPa | | | | |
| Cement | Ash | Sand | Crushed stone fraction 5...20 mm | Water | Lignosulfonates admixture | Normal hardening, days | | | Steaming | |
						28	90	180	After steaming	28 days
104	-	635	1310	160	-	13,5	17,1	17,9		
150	150	600	1240	155	0.2	18,5	20,8	24,1		
198	-	645	1340	158	0.2	15,3	19,5	20,8		
150	150	620	1280	160	0.2	19,5	27,9	30,8		
200	-	650	1350	165	0.2	17,6	24,2	23,9		
150	150	600	1240	165	-	-	-	-	20.5	22.6
200	-	650	1340	178	-	-	-	-	10.4	15.8
150	150	630	1280	160	-	-	-	-	17.5	19.1

Note: Portland Cement strength 40.2 MPa.

upon hydrolysis. Positive is an advantageous addition of ash to concrete based on Portland cement with a high mineral—silicate content is positive. It was shown that when replacing 30% of Portland cement by ash, an increase in the steamed concrete strength was 40%, while the strength of steamed concrete on cements with a lower silicate content remained at the level of concrete without ash [6].

It is important to select optimal modes of heat and moisture treatment of steamed ash-containing concrete. Selecting the modes should consider the characteristics of ash and cement used. In general, when using mixed binders containing ash or slag, high-temperature steaming is preferred. Following available experimental data, strength of ash-containing concrete steamed at 95°C was 12 ... 15% higher than that steamed at 80°C. An increase in temperature made it possible to reduce the heat treatment duration by 1... 2 hours [3].

Concrete with addition of ash exhibits a relatively higher strength at the late hardening stages. Compressive strength of concrete containing 190 and 240 kg/m^3 of cement and 30% of ash addition at 10 years, was respectively 1.44 and 1.43 times higher than that at 3 months [2]. When testing concrete pavement cores, in which 30% of cement was replaced by ash, a compressive strength of 37 MPa was observed at 3 months and 61 MPa at 9.5 years.

Table 2.7 presents data on the increase in strength of normal hardened concrete made using ash from the Ladyzhin thermal power station (Ukraine). The table shows that at 28 ... 180 days the compressive strength growth intensity of ash-containing concrete is about the same or higher than for concrete without ash.

Some researchers report that ash-containing concrete tensile and flexural strength increases intensively at long hardening. Rod and bar specimens cut from an experimental concrete masonry showed that ash-containing concrete flexural strength at 3 months is 80% and at 10 years—150% of the control concrete. Concrete with ash, as well as with other active mineral additives, has a higher ratio between tensile and compressive strengths.

For ash-containing concrete, a significant effect is achieved by using surface active substances (SAS) and hardening accelerators. Plasticizing surfactants have a deflocculating effect on highly dispersed ashes, prone to aggregation. The part of flocks in ash is 10 ... 15% and they absorb 6 ... 9 *l* of water per 100 kg. Of interest is the effect of hardening accelerating additives, in particular calcium chloride, on concrete strength. It is noted that adding 1.2 ... 1.5% of calcium chloride by the mixed binder weight enables it to increase the ash-containing concrete strength at 7 days by 18 ... 25%, and at 28 days—by 10 ... 15% [3]. Replacing part of cement by ash leads to a decrease in concrete shrinkage deformation at reduced concrete mixture water demands. The decrease in shrinkage is explained by the fact that ash adsorbs soluble alkalis from cement and forms stable insoluble aluminosilicates.

Ash, as well as other active mineral admixtures, helps to increase the cement concrete sulfate resistance. The results of a 10 year experimental program showed that concrete containing ash cement is more resistant to sea water even when compared to concrete on slag Portland cement [10].

The most significant improvement in sulfate resistance was noted for concrete on Portland cement with a high content of C_3A. The cement resistance coefficient after 6 months was 0.33, while with the addition of ash it exceeded 1, which indicates the complete absence of the investigated specimens' corrosion. The best results were noted for concrete using ashes with the highest SiO_2 + Al_2O_3 content, i.e., the most acidic in chemical composition. Using ash yields no increase in concrete durability at sulfate-magnesia and alkaline environments. Ash has little effect on concrete resistance to carbon dioxide, general acid and magnesium.

When using in concrete mixture reactive aggregates containing opal, chalcedony, silicon shells, volcanic tuffs, etc., ash can be used just if the total content of alkaline oxides in the binder by weight in terms of Na_2O is up to 0.6%. Dry ash that usually contains 1 ... 5% of alkaline oxides, can be used if it is added to almost alkaline cements. At the same time, a number of studies have shown that replacing cement by all types of ash reduces the interaction between alkalis and aggregates [7]. The upper possible limit of the total alkaline oxide content in cement-ash binder is recommended to be 1.5% [6].

Reduction in cement consumption by using ash leads to a decrease in concrete heat generation and it's heating in the initial period. Detailed studies concerning the use of ash cements in hydraulic concrete showed that heat generation in concrete on cement with 25% of ash was 15 ... 25% lower than that on cement without admixtures [8]. Analysis shows that adding a significant amount of mineral admixtures into the cements composition or directly into concrete mixtures to reduce heat generation is logical just as in those cases when no increase in water demand is caused. Such admixtures, along with blast furnace slag, include ash. Ash, like other active mineral additives, with a moderate content in the concrete mix increases concrete water resistance. At 6 months the water resistance coefficient of concrete, in which 30% of cement is replaced by ash, is 1/5, and when replacing 50% of cement, it is reduced to 1/12.

Investigation of concrete, based on cement-ash binder with 40% of ash, showed that at 28 days its water tightness class was W9, and at 180 days. - W11. Concrete based on binders with an addition of 60 and 70% of ash had W2 and W1 water tightness classes at 28 days and W8 and W3 at 180 days. Negative consequences of using ash in the concrete mixture includes a decrease in abrasion and cavitation resistance. Addition of ash to concrete is not recommended for in the autumn-winter period of work by the "thermos" method, since it slows down concrete hardening at low temperatures. During construction in areas with hot and dry climate, care for concrete containing ash should be longer than in areas with a moderate climate.

For concrete, requiring frost resistance higher than F50, as well as subjected to alternate wetting and drying, using ash is allowed after special research. In other words, the use of ash for frost-resistant concrete is allowed, but experimental justification is required. To clarify the reasons for a decrease in concrete frost resistance caused by using ash, special studies were carried out to investigate the cement stone structure [7]. It was found that when the content of ash is 30 ... 40% of the mixed binder weight, there is a sharp increase in the capillary suction and water absorption, especially at initial testing stages. Measurements showed that when replacing cement by ash, the contraction pores volume is significantly reduced. As the ash content increases, the ratio between capillary and conctional porosity changes. Capillary water absorption with addition of ash to the cement increases by about 10 ... 20% for every 10% of the ash additive [12].

The degree of concrete frost resistance, which decreases upon using ashes is different and depends on their characteristics. A significant scatter in basic physical and the mechanical properties of concrete, including frost resistance, is caused by the ash composition and properties heterogeneity.

A significant increase in the frost resistance of ash-containing concrete is achieved by using surfactants. To ensure the design frost resistance of concrete containing 20 ... 30% of ash, 0.2% of lignosulfonate by cement weight was used [12]. The strength of specimens of ordinary and ash-containing concrete after 200 freezing cycles in this case was the same. The best results were obtained

Table 2.7. Concrete strength growth rate over time.

Content, kg/m³		Compressive strength, MPa/Relative increase in strength compared to that at 28 days		
Cement	Ash	28 days	90 days	180 days
198	-	153	195/1.27	208/1.35
150	150	195	279/1.43	308/1.58
230	-	187	234/1.25	305/1.63
200	100	229	288/1.26	350/1.53
320	-	286	384/1.34	413/1.44
270	100	239	347/1.45	420/1.76
400	-	376	488/1.3	510/1.36
350	100	379	465/1.23	482/1.27

by using a complex admixture containing plasticizing (lignosulfonate) and hydrophobizing (bottoms of synthetic fatty acids) components in a ratio of 1:1 (by weight).

It is known that concrete based on composite clinker cement (slag-Portland cement and pozzolanic Portland cement) provides alkalinity of interfacial liquid in concrete, which is sufficient for steel passivation. However, such concretes are carbonized faster than Portland cement based ones. This conclusion is also true for Portland cement concrete, but with the addition of ash. Results of long-term research have shown that when 30% of cement is replaced by ash, concrete carbonization proceeds at a much greater depth, compared to that without ash. It is shown that there is a linear relationship between the W/C and the average carbonization depth. Research was carried out for a large number of specimens of ash-containing concrete with reinforcing bars, stored in natural air conditions, including in areas with harsh winters, on the sea coast, etc. At the same time, research was carried out on port facilities built 5–13 years ago using concrete with the addition of ash. It was found that the depth of carbonization in concrete is inversely proportional to its strength at the beginning of the atmospheric exposure.

Long-term tests of concrete showed that addition of ash significantly reduces the concrete creep. Testing concrete specimens during 240 days has shown that creep of concrete with ash additives was 34.5% lower than that of the control specimens. Using surfactants has a negligible effect on the creep strain of ash-containing concrete, compared to that without ash. Testing concrete with liguosulfonate admixture during 300 days shows that creep in specimens without ash additives was $59.2. \ 10^{-5}$ and in those with 20% of ash—$59.5. \ 10^{-5}$ [6].

It was revealed that ashes reduce the linear temperature expansion coefficient of concrete mortar part in an air-dry state, approaching it to the values that are typical for aggregates. So, at a temperature of 20°C, the linear expansion coefficient for ordinary mortar is 8.8, for mortar with 25% of ash and SAS admixture—5.8%, for granite—3.8%. This data shows that adding ash into concrete should increase its thermal crack resistance.

3

Self-compacting Ash-containing Concrete

One of the most effective additives for self-compacting concrete is fly ash from thermal power plants. Unlike other mineral admixtures, fly ash reduces the water demand of concrete mixtures or leaves it unchanged. However, using ash along with a reduction in cement consumption can, under certain conditions, lead to deterioration in a number of concrete technological, construction and technical properties.

The main purpose of the research presented in this chapter is to justify the optimal technological parameters for applying fly ash of various dispersion for producing concrete with improved construction and technical properties using self-compacting plasticized concrete mixtures. In the experiments, Portland cement (CEM II, 32.5), fly ash of the Ladyzhyn state district power station (Ukraine), locally available river sand and crushed granite stone with a particle size of 5 ... 20 mm were used. Polycarbocilate superplasticizer was added to the cement-ash paste and concrete mixes.

3.1 Properties of Cement-fly Ash Pastes and Fresh Concrete

Rheological properties of cement-ash paste. The available data does not allow considering the issue of the joint effect of ash content and its dispersion, which is changed at grinding, on cement-ash paste rheological properties, particularly if superplasticizers (SP) are used. To study the effects of ash on the cement paste properties, its fluidity by flow cone on a *Suttard's viscometer* and the effective viscosity measured by rotational viscometer were investigated. The studies were performed using mathematical planning of the experiment [13]. The experimental planning conditions are given in Table 3.1.

Analysis of resulting mathematical models (Eqs. 3.7 and 3.8) show that the plasticizing effect of ash and its influence on flow cone diameter D_f (y_1) and effective viscosity ln η (y_2) cannot be unambiguously determined by the

Table 3.1. Experiment planning conditions.

Factor, Type		Variation levels			Variation interval
Natural	Coded	−1	0	+1	
W/(C+A)	X_1	0.24	0.28	0.32	0.04
Part of ash D_a by weight in cement –ash mixture, %	X_2	20	40	60	20
SP superplasticizer content, % (C+A)	X_3	0.6	0.8	1.0	0.2
Specific surface of ash, S_a, cm²/g	X_4	2900	3900	4900	1000

amount of ash, added into the cement paste with SP, and largely depends on ash dispersion (Fig. 3.1). Thus, a change in ash content from 20 to 60% at specific surface area of S_a = 2900 cm²/g leads to an increase in fluidity by 2 cm (D_f value increased from 28 to 30 cm) and a decrease in effective viscosity (ln η value decreased from 2.6 to 2.3). For S_a = 3900 cm²/g an increase in ash content in the investigated interval does not lead to a change in fluidity (D_f = 27 cm) and effective viscosity (ln η = 3.2). At S_a = 4900 cm²/g a decrease in fluidity is observed (the D_f value decreases from 28 to 26 cm) and an increase in the effective viscosity (the ln η value varies from 3.2 to 3.5).

At the same time, the effect of dispersion on the investigated characteristics will become higher with an increase of the ash content in cement paste. For example, when the ash content is 20%, a change in its dispersion from 2900 to 4900 cm²/g leads to an increase in effective viscosity—the value of ln η increases from 2.6 to 3.2, the paste fluidity practically does not change (D_f = 28 cm), and at 60%—an increase in effective viscosity is even more (ln η increases from 2.3 to 3.5), but the fluidity drops from 30 to 26 cm.

$$y_1 = 27.04 + 2.86X_1 - 1.02X_4 - 2.29X_1^2 + 0.93X_1X_4$$
$$+ 0.45X_2^2 - 1.25X_2X_4 + 0.2X_3^2 + 0.7X_4^2 \qquad (3.1)$$

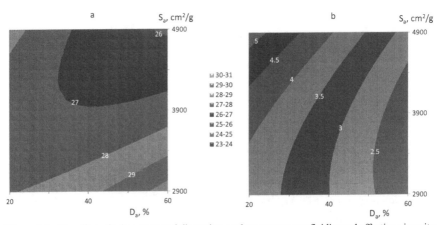

Figure 3.1. The effect of ash content and dispersion on the cement paste fluidity and effective viscosity for W/(C+A) = 0.28 and SP content of 0.8%: a—isolines of D_f, cm; b—isolines of ln η, (Pa·s).

$$y_2 = 3.24 - 1.03X_2 + 0.53X_4 + 0.48X_1^2 - 0.18X_1X_4$$
$$-0.05X_2^2 - 0.19X_2X_4 - 0.16X_3^2 + 0.3X_4^2 \tag{3.2}$$

Dependence of the cement-ash paste fluidity and effective viscosity on ash content and its dispersion is due to the change in ash water demand characterized by normal consistency (NC) with an increase in its specific surface by grinding. The following data was obtained: at $S_a = 2900$ cm^2/g the ash N.C = 16.75%; at $S_a = 3900$ cm^2/g N.C = 25.25%; at $S_a = 4900$ cm^2/g N.C = 28.25%.

Surfactant admixtures also influence the cement-ash paste fluidity and decrease in its effective viscosity. Plasticizing admixtures destroys ash and cement flocs and has a stabilizing effect, preventing the ash-cement compositions segregation as a result of sedimentation phenomena, i.e., reduce water demand and ash-cement paste viscosity. Increasing the SP content over a certain optimal value does not lead to a decrease in the initial paste viscosity.

A decisive influence on the change in the cement-ash paste fluidity and effective viscosity has a water-binder ratio W/(C+A). Increase in this ratio yields higher fluidity and lower effective viscosity. A more intensive change in the investigated characteristics was observed at low water-binder ratio values. For example, with a change in W/(C+A) from 0.24 to 0.26 at $S_a = 4900$ cm^2/g, fluidity has increased from 20.5 to 24 cm, and effective viscosity decreased from 172 to 67 Pa·s. With a change in W/(C+A) from 0.28 to 0.30 fluidity increased from 27 to 28 cm, and the effective viscosity decreased from 32 to 20 Pa·s.

Analysis of Eq. (3.1) shows that an increase in fluidity caused by an increase in the water-solid ratio is evident up to a certain value. At higher values of W/(C+A), a decrease in this parameter may occur due to the limited cement-ash paste water-holding capacity.

With an increase in the ash specific surface, the cement-ash paste water demand increases, enabling it to increase the cement-ash paste cohesion and sedimentation stability. Therefore, preliminary milling of ash is an effective method for increasing the fluidity at high W/(C+A) values. For example, with an increase in W/(C+A) from 0.30 to 0.32, the cement-ash paste flow cone at $S_a = 2900$ cm^2/g decreases from 29 to 28 cm and at $S_a = 4900$ cm^2/g there is practically no drop ($D_f = 28$ cm).

Properties of concrete mixtures. The experiments planning conditions for modeling properties of cement-ash concrete mixtures are given in Table 3.8. As a result of experimental and statistical processing, a complex of polynomial models was obtained for the following characteristics: standard cone's flow diameter, cm (y_3), superplasticizer content (SP-4), kg/m^3 (y_4) required to achieve a mixture with cone slump of 26 ... 28 cm, as well as parameters, determining uniformity and resistance of concrete mixtures to delamination: water segregation, g/l (y_5), mortar segregation,% (y_6) and tq φ (y_7). In addition, a polynomial model of the involved air volume, % (y_8) was obtained.

Table 3.2. Experiments planning conditions.

Factor, type		coded	Variation levels			Variation interval
natural		coded	−1	0	+1	interval
Water – cement ratio, W/C		X_1	0.5	0.6	0.7	0.1
Water demand, kg/m³		X_2	180	190	200	10
Part of sand in the volume of sand and crushed stone, r_s		X_3	0.34	0.41	0.48	0.07
Ash content D_a, кг/м³		X_4	50	150	250	100
Specific surface of ash, S_a, kg/m³		X_5	2900	3900	4900	1000

$$y_3 = 45.05+3.39X_1+1.39X_2-2.84X_3-0.5X_4+2.78X_5-3.07X_1^2-3.08X_2^2+1.9X_3^2-$$
$$-1.07X_4^2+0.43X_5^2+2.25X_1X_4+1.13X_2X_5+1.12X_3X_4-2.25X_3X_5 \quad (3.3)$$

$$y_4=3.28-0.07X_1-0.94X_2-1.06X_3+0.37X_4+0.7X_2^2+0.25X_3^2+0.14X_4^2-$$
$$-0.23X_5^2-0.24X_1X_2-0.25X_1X_4-0.32X_1X_5-0.17X_3X_5 ; \quad (3.4)$$

$$y_5= 0.37-0.35X_1+0.51X_2+0.75X_3-0.56X_4-0.45X_5-0.15X_1^2+0.57X_3^2-0.15X_4^2$$
$$+0.23X_5^2-0.09X_1X_2-0.09X_1X_3-0.22X_1X_5+0.41X_2X_3-0.33X_2X_4-0.28X_2X_5-$$
$$-0.32X_3X_4-0.32X_3X_5+0.2X_4X_5 \quad (3.5)$$

$$y_6=5.65+0.94X_1-2.48X_3-0.63X_4+1.28X_5-0.43X_1^2-0.43X_2^2+1.37X_3^2-1.23X_4^2-$$
$$-1.56X_5^2-0.48X_1X_2+0.46X_1X_4+0.7X_1X_5+0.56X_2X_3-0.56X_2X_4+0.81X_3X-$$
$$0,53X_3X_5 \quad (3.6)$$

$$y_7=0.23-0.08X_1-0.04X_2+0.06X_3-0.07X_5+0.12X_1^2+0.05X_2^2-0.05X_3^2+0.02X_4^2+$$
$$+0.01X_5^2-0.07X_1X_4-0.02X_3X_4 \quad (3.7)$$

$$y_8=0.86-0.24X_1-X_2+1.03X_3-0.6X_4+1.4X_5-0.2X_1^2+0.91X_2^2+0.79X_4^2+$$
$$0.05X_5^2+0.53X_1X_3-$$
$$+0.54X_1X_5-0.65X_2X_5-0.34X_3X_4+0.41X_3X_5-0.59X_4X_5 \quad (3.8)$$

For a comparative assessment of the technological factors which influence on the workability of self-compacting mixtures with almost the same cone slump, a model of cone flow diameter (y_3) can be used. The influence of the main technological factors on workability can also be monitored by analyzing the model of the required superplasticizing admixtures content (y_4).

It is known that fly ash granulometric composition, allows for the compensation of lacking grains with intermediate fineness between cement and sand. Additionally, ash particles have smooth spherical surface. These two factors increase the workability of stiff and plastic concrete mixtures. The obtained results confirmed this position for self-compacting ash-containing concrete mixtures. Moreover, higher water demand and remaining other factors unchanged leads to an increase in optimum ash consumption. At the same time,

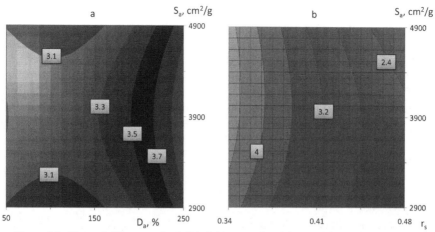

Figure 3.2. Change in SP content (kg /m³) in Self-compacting Ash-containing concrete mixtures:
a – on S_a and D_a at W/C = 0.6; W = 190 kg/m³; r_s = 0.41;
b – on S_a and r_s at W/C = 0.6; W = 190 kg/m³; D_a = 150 kg/m³.

as it follows from the models analysis (Fig. 3.2), the effect of ash content on workability should be considered taking into account its dispersion.

Analysis of models (Eqs. 3.3 and 3.4) shows that the influence of ash dispersion on the self-compacting mixture workability cannot be unambiguously estimated. In addition to ash content (D_a), the influence of the ash specific surface depends on W/C and r_s of concrete mixtures. Analysis of mathematical models (Eqs. 3.3 and 3.4) demonstrates that an increase in the fly ash specific surface at its optimal content and high W/C values increases the self-compacting concrete mixture workability. With a decrease in W/C, the positive effect of ash grinding decreases. An increase in ash dispersion at optimal content and maximum sand part (in the investigated intervals) yields an increase in the self-compacting mixture workability and at a minimum sand part, workability deteriorates (Fig. 3.2b).

This fact is due to a change in the particle size distribution of the ash during its grinding. When grinding ash, first of all, the content of large particles sharply decreases, which, with a reduced proportion of sand, adversely affects the overall aggregates granulometry, while an increased proportion of sand, on the contrary, contributes to creating continuous granulometry that is most favorable for achieving maximum workability. The most important properties, characterizing self-compacting concrete mixtures quality, are water segregation and delamination.

Analysis of mathematical model (Eq. 3.5) shows that efficiency of fly ash addition increases with an increase in its content and with a decrease in W/C. The latter is the dominant factor, determining water segregation. So, at ash content of 90 kg and W/C = 0.5, the water segregation value is 0.25 g/l, and with an increase of W/C to 0.7, water segregation increases to 1.45 g/l (Fig. 3.3). At W/C = 0.7, an increase in D_a from 50 to 150 kg/m³ leads to a decrease in water segregation from 1.85 g/*l* to 1.05 g/*l*. A further increase in D_a to 250 kg/m³ allows us to reduce the water segregation to 0.45 g/*l*.

Along with an increase in ash content, an increase in its dispersion also has a positive effect on self-compacting ash-containing concrete water segregation (Fig. 3.3). For example, when the ash content is 250 kg and its specific surface is 2900 cm²/g, the water segregation is 0.45 g/*l*, and when the specific surface is increased to 3900 cm²/g, the water segregation value is reduced to 0.15 g/*l*. Thus, the water segregation sensitivity to ash content makes possible to use it at a rational content of 150 ... 200 kg/m³ as a water retaining admixture in self-compacting concrete mixtures after increasing their W/C at additional grinding. Analysis of Eq. (3.11) shows that the positive effect of increasing the ash content on water segregation becomes even higher with an increase in the part of sand in the aggregates mixture.

A special feature of self-compacting mixtures is the possibility of internal and external segregation. The first is due to the action of gravity. The segregation decreases as the mortar part viscosity increases and the size of the aggregate grains decreases. The second occurs as a result of insufficient adhesion between coarse aggregate and mortar component, which may be due to the excessively high viscosity of the latter or high crushed stone or gravel content.

Based on analysis of mathematical models (Eqs. 3.6 and 3.7), it is evident that the addition of fly ash has a positive effect on the decrease in the concrete mixture segregation level. At the same time, the extreme nature of this effect (Fig. 3.4) can be explained by the fact that the best granulometry of the mixture corresponds to the optimal ash content. The optimum ash content, required to ensure minimum mortar segregation, exceeds that, corresponding to the maximum value of tg $\varphi = (30 - CS)/0.5 \, D_f$, where CS is the cone slump and D_f is the diameter of the cone flow (Fig. 3.4 a) With a reduction in ash content, concrete mixtures without stratification can be obtained by increasing the sand

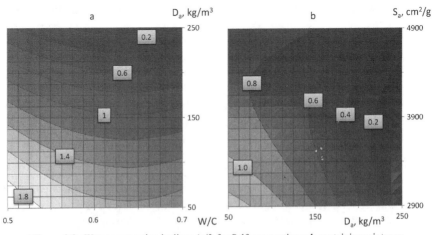

Figure 3.3. Water segregation isolines (g/*l*) for Self-compacting ash containing mixtures:
a – W = 190 kg/m³ ; r_s = 0.41; S_a = 4900 cm²/g;
b – W/C = 0.6; W = 190 kg/m³ ; r_s = 0.41.

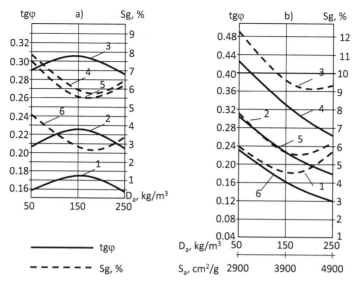

Figure 3.4. The Effect of ash content and dispersion on tgφ and mortar segregation (S_g) of Self-compacting Ash-containing Concrete Mixtures:
a – effect of D_a on the value of tgφ and mortar segregation at W/C = 0.6; W = 190 kg/m³; r_s = 0.41 and S_a: 1, 4 - 4900; 2, 5 - 3900; 3, 6 - 2900 cm²/g;
b – effect of D_a on the value of tgφ and mortar segregation at В/Ц = 0.6; W = 190 kg/m³; S_3 =3900 cm²/g; r_s : 1 - 0.48; 2 - 0.41; 3 - 0.34; effect of S_a on the value of tgφ at W/C = 0.6; W = 190 kgr/m³; r_s = 0.41; D_a = 150 kg/m³ and W/C: 4 - 0.5; 5 - 0.6.

part (Fig. 3.4). An increase in ash dispersion by grinding leads to a significant increase in the cement-ash paste viscosity with addition of SP, which determines the increase in viscosity of mortar component of concrete.

Excessively high mortar component viscosity is due to an increase in the self-compacting ash-containing concrete segregation with an increase in ash dispersion. However, it should be noted that when selecting the optimal compositions of such concrete, the segregation level does not exceed the permissible limits (tg φ ≥ 0.2 and mortar segregation of 5%).

As it can be seen from analysis of mathematical models (Eqs. 3.5, 3.6 and 3.7), the optimal ash content and its dispersion are different for self-compacting ash-containing concrete water segregation and segregation parameters. For example, an increase in ash content in a concrete mixture from 150 to 250 kg/m³ at W/C = 0.6 reduces the value of tg φ (Fig. 3.4), while at the same time having a positive effect on water segregation (Fig. 3.3). Therefore, the choice of optimal ash consumption and its dispersion must be made in a compromise zone, allowing and obtaining concrete mixtures that satisfy all the above-mentioned requirements.

Air entraining in self-compacting concrete mixtures is one of the main factors determining the strength, uniformity and durability of concrete. It is known that each percent of involved air causes a 5 ... 10%.decrease in concrete strength. Air is entrained at mixing, unloading and casting of the concrete mix and depends on its composition, characteristics of its components and SAS content. Using SP admixtures in self-compacting concrete mixtures composition allows increasing air entraining up to 3%. However, such mixtures have a low air content due to its rapid loss caused by low viscosity.

Using fly ash after additional grinding increases the cement systems viscosity, which in turn contributes to keeping the entrained air in self-compacting mixtures with SP. At the same time, fly ash in concrete performs as active mineral admixture, which predetermines a decrease in its porosity and the formation of a more dense concrete structure. Analysis of the mathematical model given in Eq. (3.8) confirms these theoretical premises.

An increase in ash content in the investigated intervals leads to a decrease in air involvement in self-compacting concrete mixtures. Increasing the ash dispersion, on the contrary, contributes to increase in the amount of the involved air. For example, at $W/C = 0.5$; $W = 180$ kg/m³; $r_s = 0.41$; $S_a = 3900$ cm²/g and an ash content of 50 kg/m³, the entrained air volume is 3.8%, at ash content of 150 kg - 2.6%, and at an ash content of 250 kg - 2.15%. For the same W/C and W values, $r_s = 0.48$ and $D_a = 250$ kg/m³, an increase in S_a from 2900 to 4900 cm²/g leads to a corresponding increase of entrained air volume from 0.9% to 5.6%, i.e., to such a volume that should provide a sufficiently high frost resistance of concrete. Therefore, to achieve a certain air entraining by increasing the ash content, its dispersion should be accordingly increased. For example, at $D_a = 50$ kg/m³ to ensure air entrainment of 1.05%, the minimum ash dispersion should be $S_a = 3500$ cm²/g, at $D_a = 150$ kg/m³ – $S_a = 4050$ cm²/g, and at $D_a = 250$ kg/m³ the minimum ash dispersion should exceed 4400 cm²/g. The choice of ash content providing the required air entrainment should be carried out considering S_a and also r_s. At the same time, the choice of S_a depends on interaction between all of the investigated factors.

The positive effect of increase in S_a on air entraining is more pronounced with an increase in r_s. An increase in S_a from 2900 to 4900 cm²/g at $r_s = 0.34$ makes it possible to achieve air entraining of 1.77% in a concrete mixture containing almost no air, and at $r_s = 0.48$ the air entraining increases from 0.71 to 4.42%. Thus, ensuring the necessary air entrainment as well as other properties of self-compacting concrete mixtures with fly ash and superplasticizing admixtures that can be achieved by controlling the ash content and dispersion.

3.2 Strength Properties of Self-compacting Ash Containing Concrete

To study the dependencies that determine the self-compacting cement-ash concrete strength properties, two series of experiments were carried out: the first—in accordance with the planning conditions given in Table 3.2 and the second in Table 3.3. The second series of experiments was performed using ash with constant dispersion and changing the sand fineness modulus. The integral aggregates composition was characterized in this case by two volumetric ratios:

r_1—part of crushed stone in the total volume of crushed stone, sand and ash, and r_2—part of ash in the total volume of ash and sand. Experimental-statistical models of concrete strength properties obtained in the first and second series of experiments are given in Table 3.4 and 3.5 respectively.

For analysis of strength models, iso-level diagrams were constructed (Fig. 3.5). In accordance with these diagrams, the range of possible ash content values narrows with an increase in concrete strength or water-cement ratio at a constant strength. For example, to obtain ash-containing concrete with a strength of 26 MPa at a 28 days at W/C = 0.56, the ash content can range from 78 to 250 kg, and for concrete with a strength of 30 MPa, with the same water-cement ratio, the possible ash content values are in the range of 118–222 kg. To achieve the same concrete mixture workability when changing the ash content, it is necessary to adjust the superplasticizer content.

Analysis of Fig. 3.5 shows possibility of choosing the optimal ash content and dispersion. If the minimum possible ash content is taken as the optimal, then with an increase in W/C it increases. However, for practical purposes the optimum ash content can be assigned uniformly over the entire range of concrete strength. This content be found analytically from Eq. (3.9) by defining the partial derivative by X_4: $dy_9/dx_4 = 2.52 - 12.4X_4$. For accepted conditions $X_{4\,opt.}$ 0.2, i.e., $D_{a\,opt.} = 170$ kg/m^3.

Table 3.3. Experiments planning conditions.

Factor, type		Variation levels			Variation interval
natural	coded	−1	0	+1	
Water-cement ratio, W/C	X_1	0.5	0.6	0.7	0.1
Water demand, W	X_2	180	190	200	10
Part of crushed stone in the total volume of crushed stone, sand and ash, r_1	X_6	0.45	0.53	0.61	0.08
Part of ash in in the total volume of ash and sand, r_2	X_7	0.1	0.2	0.3	0.1
Sand fineness modulus, M_f	X_8	1.4	2.4	3.4	1

Table 3.4. Experimental-statistical models of concrete strength properties (1st series).

Indicator	Model
Compressive strength at 28 days of normal hardening (f_c), MPa	$y_9 = 26.1 - 4.57X_1 - 0.65X_3 + 2.52X_4 - 2.17X_5 + 1.78X_1^2 + 3.68X_2^2 + 2.43X_3^3 - 6.21X_4^2 - 4.71X_5^2 - 1.74X_1X_3 + 1.28X_2X_5 - 0.63X_3X_5$ (3.9)
Cement efficiency coefficient $K_e = f_c/C$	$y_{10} = 0.82 - 0.003X_2 + 0.008X_4 - 0.007X_5 + 0.012X_2^2 + 0.007X_3^2 - 0.019X_4^2 - 0.015X_5^2 - 0.02X_1^2 - 0.005X_1X_2 - 0.0027X_1X_5 + 0.0046X_2X_5 - 0.002X_3X_5$ (3.10)
Concrete strength variation coefficient, (V_s),%	$y_{11} = 4.33 + 1.63X_1 - 2.3X_2 + 0.08X_3 + 1.13X_4 - 0,6X_5 - 1.07X_1^2 + 2.03X_2^2 + 7.34X_3^2 + 0.48X_4^2 - 2.97X_5^2 - 2.03X_1X_2 + 0.64X_1X_3 + 0.84X_1X_4 + 0.21X_1X_5 + 2.49X_2X_3 + 1.13X_3X_4 - 1.18X_3X_5 + 0,85X_4X_5$ (3.11)

Table 3.5. Experimental-statistical Models of Concrete Strength Properties (2nd series).

Indicator	Model
Compressive strength at 28 days of normal hardening (f_c), MPa	$y_{12} = 28.60 - 5.65X_1 + 2.75X_7 + 2.61X_1^2 - 0.53X_2^2 - 2.7X_6^2 - 3.0X_7^2 - 6.5X_8^2 + 1.93X_1X_7 + 0.98X_2X_6$　(3.12)
Same, at 90 days	$y_{13} = 35.44 - 6.26X_1 + 3.54X_4 + 2.67X_5 + 6.53X_1^2 - 3.43X_2^2 + 3.3X_3^2 - 2.97X_4^2 - 4.4X_5^2$　(3.13)
Same, at 180 days	$y_{14} = 45.9 - 0.7X_1 + 2.55X_2 + 2.76X_4 + 2.32X_5 + 0.22X_1^2 + 1.42X_2^2 + 5.92X_3^2 - 7.34X_4^2 - 7.73$　(3.14)
Concrete compressive strength, MPa: 4 *h* after steaming by (2)+3+6+2 mode at 80°	$y_{15} = 26.4 - 4.65X_1 + 3.6X_7 + 1.6X_1^2 - 0.99X_2^2 - 0.19X_6^2 - 1.65X_7^2 - 3.8X_8^2 + 1.35X_1X_7 + 1.06X_2X_6$　(3.15)
Same, 28 days after steaming	$y_{16} = 29.7 - 7.09X_1 + 3.15X_7 + 3.3X_1^2 - 1.09X_2^2 - 0.3X_6^2 - 1.6X_7^2 - 4.5X_8^2 + 1.25X_1X_7 + 0.67X_2X_6$　(3.16)
Concrete splitting tensile strength at normal hardening , MPa: At 28 days	$y_{17} = 2.24 - 0.28X_1 + 0.27X_4 + 0.15X_1^2 + 0.04X_2^2 + 0.15X_3^2 - 0.26X_4^2 - 0.3X_5^2 - 0.18X_3X_4$　(3.17)
Same, at 90 days	$y_{18} = 2.6 - 0.4X_1 + 0.19X_4 + 0.29X_5 + 0.51X_1^2 + 0.07X_2^2 - 0.06X_3^2 - 0.27X_4^2 - 0.23X_5^2$　(3.18)
Same, at 180 days	$y_{19} = 2.9 - 0.45X_1 + 0.19X_4 + 0.18X_5 + 0.26X_1^2 + 0.04X_2^2 + 0.52X_3^2 - 0.49X_3^2 - 0.32X_5^2 - 0.27X_1X_2 + 0.19X_1X_3$　(3.19)
Concrete splitting tensile strength, MPa: 4 *h* after steaming	$y_{20} = 1.74 - 0.23X_1 + 0.33X_4 + 0.11X_1^2 - 0.08X_2^2 - 0.04X_3^2 + 0.15X_4^2 - 0.18X_5^2 - 0.2X_3X_5$　(3.20)
Same, 28 days after steaming	$y_{21} = 2.24 - 0.33X_1 + 0.26X_4 + 0.14X_1^2 + 0.2X_2^2 + 0.15X_3^2 - 0.19X_4^2 - 0.46X_5^2$　(3.21)

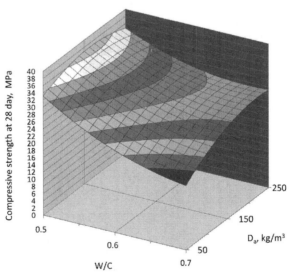

Figure 3.5. Strength of self-compacting ash-containing concrete response surface for W = 190 kg/m³, $r_s = 0.53$.

Most of the works noted [2, 8] increase in concrete strength as ash dispersion becomes higher. However, sometimes with a constant water-cement ratio, an increase in ash dispersion does not lead to additional strength growth [14]. Obviously, the nature of ash dispersion effect on concrete and mortar strength depends on a combination of factors: initial fluidity, water-binder ratio, aggregates porosity, type of ash, concrete hardening conditions, etc. On one hand, an increase in ash dispersion by grinding should lead to higher chemical activity, which has a positive effect on concrete strength growth. On the other hand, an increase in ash dispersion causes a higher air content of concrete mixture and mortar component viscosity. The latter, when it exceeds the permissible values, leads to external segregation. It should be noted that concrete strength depends on granulometry and content of its components. Excessive increase in fine ash fractions content can lead to decompression of self-compacting concrete.

Analysis of concrete strength model (Eq. 3.9) shows that, at constant W/C and ash content, the effect of its dispersion on self-compacting concrete strength has an extreme character (Fig. 3.6). In contrast to the optimal ash content, its optimum dispersion depends on water demand and the part of sand in the aggregate mixture. The analytical expression for an optimal specific surface under the accepted modeling conditions has the following form

$$X_{s\,opt} = (1.28X_2 - 0.63X_3 - 2.17)/9.42 \qquad (3.22)$$

However, a change in r_s and water demand have no significant effect on $S_{a\,opt}$. For example, when

$X_2 = 0$ (W $= 190$ kg/m^3); $X_3 = 0$ ($r_s = 0.41$); $X_s = -0.23$($S_{a\,opt} = 3670$ cm^2/g) at $X_2 = 1$(W $= 200$ kg/m^3); $X_3 = -1$($r_s = 0.34$); $X_s = 0$($S_{a\,opt} = 3870$ cm^2/g) and for $X_2 = -1$ (W $= 180$ kg/m^3); $X_3 = +1$($r_s = 0.48$); $X_s = -0.43$ ($S_{a\,opt} = 3470$ cm^2/g).

Therefore, the optimum ash dispersion, similar its optimal content, can be set constant for a wide range of self-compacting concrete strength. The range of values for $S_{a\,opt}$ is 3500 ... 4000 cm^2/g.

Cement consumption in self-compacting ash-containing concrete with optimal ash content values and dispersion is significantly lower than in conventional plasticized self-compacting concrete. For example, to achieve self-compacting concrete strength of $f_c = 31$ MPa at 28 days for W $= 180$ kg/m^3; W/C $= 0.5$ and $r_s = 0.43$ using SP admixture without ash, the required cement consumption is 360 kg. The same strength of self-compacting ash-containing concrete can be reached at cement consumption of 300 kg and ash content of 150 kg/m^3 at specific surface area of 2900 cm^2/g. Consequently, cement saving is about 60 kg per 1 m^3 of self-compacting concrete. The obtained experimental and calculated data indicates possibility of reducing cement consumption by 17 ... 20% while ensuring the specified strength by using the optimal amount of fly ash in self-compacting concrete compositions. Grinding fly ash to optimum specific surface allows to achieving additional cement savings of 5 ... 8%.

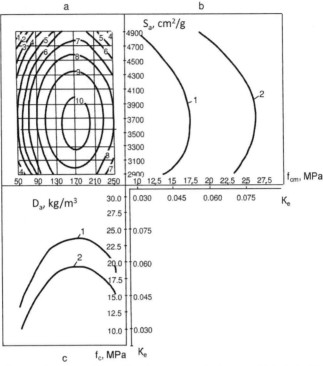

Figure 3.6. Combined diagram of strength and cement use efficiency coefficient of self-compacting ash-containing concrete:

a – iso-lines for strength (table 3.6) at W/C = 0.6; W = 190 kg/m³, r_s = 0.41;

b – concrete strength distribution and cement use efficiency coefficient at: 1 – D_a = 50 kg/m³; 2 – D_a = 150 kg/m³;

c – concrete strength distribution and cement use efficiency coefficient at: 1 – S_a = 290 m²/kg; 2 – S_a = 490 m²/kg.

To optimize the composition of ash-containing concrete, a technique, based on efficiency criterion, characterized by a corresponding efficiency cement use coefficient (K_e) (specific strength per unit of cement consumption).was proposed. Analysis of mathematical model (Eq. 3.11) and its graphical representation (Fig. 3.6), depending on variables D_a and S_a, shows that the influence of both ash content and dispersion on this criterion have an extreme character. It is known that for moderately self-compacting concrete at 28 days K_e is 0.065 ... 0 085. Adding ash into the composition of such concrete contributes to an increase K_e to 0.067 ... 0.12. The obtained results prove the feasibility of achieving the same maximum K_e values of self-compacting ash-containing concrete with SP admixture such as for moderate flowable concrete. Evaluation of concrete properties uniformity was carried out according to results of testing twelve cubic specimen with an edge of 10 cm for the investigated compositions. As a quantitative uniformity characteristic was used concrete strength variations coefficient V_s.

$$V_m = (S_{cs}/f_{cm})\ 100 \qquad (3.23)$$

where f_{cm} is the average concrete strength, MPa; S_{cs} is the mean square deviation of concrete strength, MPa.

After the experimental part was completed, the required number of n concrete samples was determined to obtain reliable results:

$$n = V_m^2 t^2/E^2 \qquad (3.24)$$

where t is the criterion of Student, found with corresponding confidence probability and number of degrees of freedom;

E is accuracy indicator (5%).

Based on the actual average coefficient of concrete strength variation, calculated according to (3.23), the required number of samples was at least ten.

Analysis of the model for concrete strength variation coefficient (Eq. 3.11) shows that the effect of ash on uniformity of self-compacting concrete strength has a complex character. Uniformity depends on the ash content and its dispersion, but on interaction of these and all other investigated technological factors.

Using unmilled fly ash into flue concrete with SP admixture (Fig. 3.7) helps to increase the concrete properties uniformity with an increase in ash content. An increase in ash dispersion S_a to 4000 cm²/g leads to an increase in V_m. The nature of the effect on concrete strength uniformity at further increase in S_a ambiguous depends on D_a. For example, with a minimum (within the investigated interval) ash content ($D_3 = 50$ kg/m³), a change in specific surface from 4000 to 4900 cm²/g leads to a decrease in V_m from 4.92% to 0.78%, with a maximum ($D_a = 250$ kg/m³)—a change in S_a from 4000 to 4500 cm²/g does not change the variation coefficient value and just a little later its decrease is observed. At $S_a = 4900$ cm²/g the variation coefficient is already 3.54%.

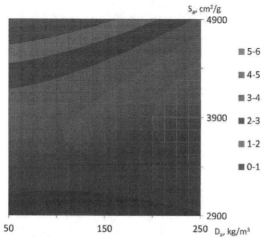

Figure 3.7. Iso-lines of strength variation coefficient (%) for self-compacting ash containing concrete at W/C = 0.6; W = 190 kg/m³ and $r_s = 0.41$.

To achieve the maximum possible uniformity of self-compacting concrete strength at initial ash dispersion of (S_a = 2900 cm²/g), its content should be maximum (D_a = 250 kg/m³), and vice versa, for maximum ash specific surface of (S_a = 4900 cm²/g) its content should be minimum (D_a = 50 kg/m³).

Thus, the optimal ash content and dispersion obtained from the condition of ensuring the maximum uniformity of self-compacting concrete strength are different from those corresponding to the maximum strength condition. When selecting the composition of such concrete, it is advisable to select the ash content and dispersion in a certain zone of optimal values, taking into account all the requirements for concrete mix and concrete.

From the analytical mathematical models calculation which that with an increase in W/C, the required part of ash (r_2) in the total volume of ash and sand increases. At the same time, increase in strength is insignificant (up to 1 MPa), which predetermines the possibility of choosing a constant ash share for concrete at certain age. In turn, if the ash part is constant, its content is practically independent of W/C. To find the part of ash, it can be considered constant. For steamed concrete at 28 days, r_2 is higher than for the normally hardened one of the same age, which is the result of intensification in the interaction between ash with $Ca(OH)_2$ under conditions of heat and moisture processing.

The optimal ash content in self-compacting concrete decreases with age. At high W/C at 28 days it can increase and reach a maximum value of 235 kg/m³ (r_2 = 0.276). At 180 days the optimum ash content is already in the range of 165 ... 198 kg/m³ (r_2 = 0.204 ... 0.233). The interaction reaction of ash with $Ca(OH)_2$ weakens with age and is covered by the negative effect of a significant dilution of binder with an low-active component, leading to a decrease in optimal ash content. Thus, when selecting compositions based on ensuring strength conditions, the optimal ash content of natural hardening self-compacting concrete should be in the range of 150 ... 200 kg/m³.

It is known that with an increase in the compressive strength, the tensile strength also increases, but to a lower extent. Addition of fly ash to self-compacting concrete, contributes to a significant increase in compressive strength, however, it does not have a valuable effect on tensile strength within the entire investigated time interval. This can be explained by the fact that using fly ash contributes to creation of additional crystallization centers that provides increased compressive strength and more favorable conditions for crystal growth, which in turn contributes to the appearance of stresses, negatively affecting tension. Analysis of models (3.17 ... 3.21 in Table 3.5) allows choosing the optimal ash part to ensure the splitting tensile strength of self-compacting concrete. Similarly to the optimal ash part, from the compressive strength condition, its choice depends basically on W/C.

The nature of concrete hardening conditions influence along with the age, factor on the choice of r_2 to ensure that the compressive and tensile strengths are identical. At the same time, the ash part value varies depending on what strength indicator is dominant, which once again confirms the need to select the optimal ash part (content) from conditions for ensuring a set of concrete properties.

The choice of crushed stone part (r_1) depends on the concrete mixture water content. The effect of change in r_1 on compressive strength is significant only for normally hardened concrete (Fig. 3.8).

Concrete strength is highly dependent on sand M_f (Fig. 3.9). For example, replacing sand with $M_f = 1.4$ by that with $M_f = 2.4$ can increase the concrete compressive strength by 15 ... 25%. The influence of M_f on strength depends on the self-compacting concrete age. For splitting tensile strength, the influence of M_f with age is less noticeable.

Compressive and splitting tensile strength of self-compacting ash-containing concrete with optimal technological factors reach the following maximum values (MPa); after steam curing—37.8 and 2.76; 28 days after steam curing—43.3 and 3.5; 28 days of normal hardening—45.5 and 3.1; 90 days of normal hardening—51.8 and 4.2; 180 days of normal hardening—65.8 and 5.1.

The increase in strength over time is consistent with the structural formation features of self-compacting ash-containing concrete. High values of compressive strength growth coefficient are the result of structure formation process intensification. Due to the ash ability to bind free lime, formed during Portland cement hardening over time and leading to the creation of additional crystallization centers, a relatively intensive increase in self-compacting concrete strength is observed. The obtained results indicate the possibility of achieving the strength growth coefficient values at 180 days exceeding 1.6 ... 1.7 for W/C = 0.6; for W/C = 0.5 its value may fall below 1.5. For ordinary

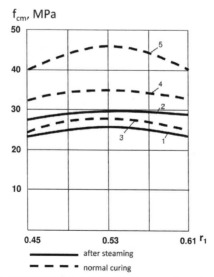

Figure 3.8. Effect of r_1 on compressive strength of self-compacting ash containing concrete at:
W/C = 0.6; W = 190 kg/m³; $r_2 = 0.2$; $M_f = 2.4$;
1 – after steam curing; 2, 3 - 28 days; 4 - 90 days.; 5 - 180 days.

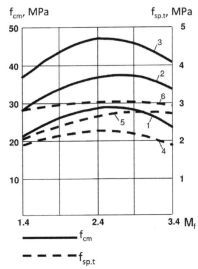

Figure 3.9. Effect of M_f on compressive (f_c) and splitting tensile strength ($f_{sp.t}$) of self-compacting ash containing concrete at W/C = 0.6; W = 190 kg/m³; r_1 = 0.53; r_2 = 0.2: 1, 4 - 28 days; 2, 5 - 90 days; 3, 6 - 180 days.

concrete after steaming, depending on its cement composition, it is possible to achieve 50 ... 80% strength at 28 days of normal hardening conditions. Higher concrete strength can be obtained only using particularly stiff mixtures at long steaming duration.

The results, obtained in our work, demonstrate that the compressive strength of self-compacting ash-containing concrete after steaming can reach more than 90% of strength at 28 days. Moreover, increase in ash dispersion (by milling) allows achieving higher strength values after steam curing in relation to that at 28 days.

3.3 Durability of Self-compacting Concrete Containing Fly Ash

<u>Crack resistance</u>. Crack resistance of concrete depends on such factors such as thermal expansion, creep, changes in concrete elastic-plastic characteristics vs. temperature, etc. Therefore, selection of corresponding criterion and evaluation of concrete crack resistance is a very difficult task. Reducing the concrete cracking risk with an addition of ash is associated with a decrease in heat hydration. However, according to thermal crack resistance criterion, it is difficult to obtain the relative crack resistance (especially because the need in considering the scale factor influence when passing from specimen to structure). For this purpose, it is most appropriate to use another criterion—a ratio of tensile to compressive strength $(f_t/f_c) \cdot 100$. An increase in this ratio is considered more favorable from the viewpoint of crack resistance and it varies in time from 15 ... 18% at 3 days to 6 ... 7% at 180 days.

Analysis of mathematical models (3.12 ... 3.21) enables us to assume that for self-compacting ash-containing concrete with the addition of SP, hardening in natural conditions, the ratio f_t/f_c decreases with an increase in the time interval and at 180 days it reaches 6 ... 7%, which is typical for most concretes. The decrease in the f_t/f_c ratio in time reflects the features of the cement stone structure formation processes, concrete composition and its hardening conditions [15, 16]. Provisions known for ordinary cement concrete can be fully attributed to self-compacting ash-containing concrete. For example, a change in the crushed stone part in the aggregates mixture makes it possible to stabilize the f_t/f_c ratio in time interval of 90 ... 180 days. It should be noted that with an increase in self-compacting ash-containing concrete age, the number of factors and their interactions that affect the f_t/f_c ratio decreases.

Regulation of ash part in the fine aggregate mixture (within the investigated interval) has a negligible effect on the nature of f_t/f_c change in time and, consequently, on the nature of the crack formation process. Minimization of the ash content enables it to slow down the strength indicators ratio falling process just in the 28 ... 90 days interval. The f_t/f_c ratio for self-compacting ash-containing steamed concretes, were investigated both immediately after steaming and at 28 days. The strength ratio in this age interval, as a rule, increases. The influence of the investigated technological factors on f_t/f_c for self-compacting steamed ash-containing concrete, it is more complex than for normal hardening one, as the number of factors and their interactions affecting this ratio increases. In contrast to normally hardened concrete, the choice of optimal ash part in fine aggregate mixture of self-compacting ash-containing steamed concrete is important for the f_t/f_c ratio (tensile to compressive strength ratio). Achievement of the maximum ash part does not enable to reach the maximum f_t/f_c value both immediately after steaming and at 28 days if the influence of interaction with other technological factors is not considered.

<u>Watertightness.</u> Watertightness of concrete depends mainly on its porous structure features. If the total porosity has the main effect on concrete strength, then water tightness is a function of open through porosity. The main ways of water penetration into concrete are pores of sedimentary origin, formation of which is most typical for self-compacting concrete. Sedimentation depends mainly on the cement paste viscosity and the main sedimentation characteristic of the concrete mix is its water segregation. As it was shown earlier, addition of fly ash has a significant effect on the plasticized cement paste viscosity and self-compacting concrete mixture water segregation. Due to addition of fly ash into the concrete mixture, the number of micropores is reduced and open pores are clogged. The mathematical model of self-compacting ash-containing concrete watertightness (y_{22}, MPa), obtained based on experimental data (planning conditions given in Table 3.9), confirms the theoretical assumption:

$$y_{22} = 0.93 - 0.15X_1 + 0.06X_6 + 0.02X_1^2 - 0.04X_6^2 - 0.08X_7^2 - 0.03X_8^2 - 0.03X_1X_2 -$$
$$0.03X_1X_8 - 0.07X_2X_6 + 0.03X_2X_7 + 0.03X_6X_7 + 0.05X_7X_8 \qquad (3.25)$$

As for traditional concrete, the water-cement ratio is the main factor, determining the watertightness of self-compacting ash-containing concrete. A sharp decrease in self-compacting ash-containing concrete watertightness at W/C less than 0.6 ... 0.7 (Fig. 3.3) can be explained by a decrease in the internal water segregation in concrete mixtures. Increasing the concrete watertightness is associated with the choice of the optimal ash content (Fig. 3.10). Selecting the value of r_2 depends on W, M_f and r_1. The analytical expression of r_2 for watertightness is:

$$X_{7\,opt} = (0.03X_2 + 0.03X_6 + 0.05X_8)\, 0.16 \qquad (3.26)$$

By selecting the optimal values of factors and their combinations, self-compacting ash-containing concrete water tightness at 28 days can reach 1.2 MPa. With an increase in concrete age watertightness continuously becomes higher (Fig. 3.11) as a result of irreversible changes in the pore space structure and cement stone solid phase volume growth. At sufficient hardening mode moisture, the increase in concrete watertightness at the late hardening stages is much greater than the relative increase in compressive strength. At the same time, the most intensive increase in water tightness was noted for concrete with high W/C values and less for low W/C.

In order to study the effect of ash fineness on permeability, porosity character of self-compacting concrete specimens with addition of fly ash of different dispersion (obtained by grinding) was investigated. The parameters of the self-compacting concrete specimens pore structure are presented in Table 3.6.

Adding unmilled ash slightly reduces the number of open capillary pores in self-compacting concrete. With an increase in ash dispersion to $S_3 = 3900$ cm²/g, the total number of capillary pores also remains practically unchanged. However, using original and even more milled ash leads to a significant decrease in the equivalent pore radius and increase in the pores specific surface area, which determines a decrease in concrete permeability.

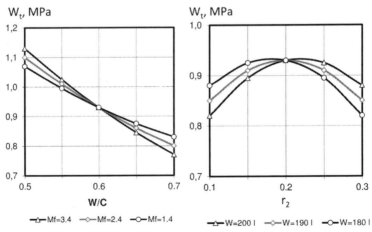

Figure 3.10. Watertightness (W_t) of flow ash containing concrete.

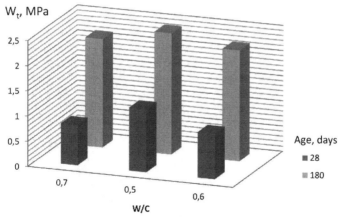

Figure 3.11. Change in watertightness (W_t) of self-compacting ash containing concrete in time.

Table 3.6. Parameters of pore structure of concrete.

No.	S_a, cm²/g	Open capillary porosity	Equivalent hydraulic radius, cm	Capillary specific surface, m²/cm³
1	2500	0.137	$2.67 \cdot 10^{-6}$	5.14
2	2900	0.132	$2.23 \cdot 10^{-6}$	5.9
3	3900	0.134	$1.05 \cdot 10^{-6}$	12.7
4	4900	0.123	$1.15 \cdot 10^{-6}$	10.67

Note. Cement consumption for composition 1 is 410 kg/m³, for other compositions—360 kg/m³; water demand for all compositions—185 kg/m³, SP content is 0.8% by cement weight.

An increase ash dispersion to $S_a = 4900$ cm²/g, despite a decrease in the total number of open capillary pores, leads no further decrease in the hydraulic radius and increase in open capillary pores specific surface, which explains no additional increase in concrete watertightness compared to concrete containing ash with $S_a = 3900$ cm²/g. Thus, to increase watertightness of self-compacting ash-containing concrete with SP, it is advisable and sufficient to grind ash to $S_3 = 3900 \dots 4000$ cm²/g.

Frost resistance. Concrete frost resistance is determined by the nature of its pore structure. Capillary pores are the main defect in compact concrete structure, reducing its frost resistance. Concrete with capillary pore content less than $5 \dots 7\%$ is frost-resistant. In addition, pore sizes have a significant impact on frost resistance—pores with a size of more than 10^{-5} cm negatively affects it. The obtained results (Fig. 3.12) indicate higher open porosity (more than 10%) of self-compacting concrete and explain a possible decrease in their frost resistance in comparison with traditional concrete:

Composition No. According to Table. 3.7	1	2	3	4
Critical number of freezing and thawing cycles	240	310	350	360

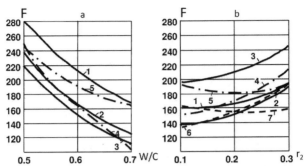

Figure 3.12: Influence of the investigated technological factors on frost resistance (F) of Self-Compacting Ash Containing Concrete (cycles);
$a - r_1 = 0.53$; $M_f = 2.4$; $r_2 = 0.2$ and W: 1 - 180 kg/m³; 2 - 190 kg/m³; 3 - 200 kg/m³;
W = 190 kg/m³; $r_1 = 0.53$; $M_f = 2.4$ and r_2: 4 – 0.1; 5 – 0.3;
b – W/C = 0.6; $r_1 = 0.53$; MK = 2.4 and W: 1 - 190 kg/m³; 2 - 200 kg/m³; 3 - 180 kg/m³;
W/C = 0.6; B = 190 kg/m³; $r_1 = 0.53$ and M_f: 4 – 3.4; 5 – 1.4;
W/C = 0.6; B = 190 kg/m³; $M_f = 2.4$ and r_1: 6 – 0.61; 7 – 0.45.

Addition of ash, despite a slight decrease in the total number of open capillary pores, leads to an increase of frost resistance in self-compacting concrete, which is a result of the decrease in pore size. The specific surface of open capillary pores increases, and the equivalent hydraulic radius decreases (Table 3.7).

Using ash grinding to S_a = 3900 cm²/g further contributes to a decrease in the size of open capillary pores and, as a result, to increase in the frost resistance of self-compacting concrete with SP. Grinding ash to S_a = 4900 cm²/g, despite the reduction in the total number of open capillary pores (Table 3.12), leads to no further reduction in their size and slightly increases the self-compacting concrete frost resistance. Thus, an increase in self-compacting concrete frost resistance with ash addition is decisively associated with a decrease in the size of concrete capillary pores, and not their quantity. To increase the self-compacting ash-containing concrete frost resistance, it is advisable and sufficient to grind ash to S_a = 3900 ... 4000 cm²/g.

An experimental-statistical model of self-compacting ash-containing concrete frost resistance in cycles was obtained in accordance with the planning conditions given in Table 3.3:

$$y_{23} = 165 - 57X_1 - 30X_2 + 16X_7 - 6X_8 + 15X_1^2 + 15X_2^2 - 11X_6^2 + 10X_7^2 - 18X_1X_6 +$$
$$17X_1X_7 + 18X_2X_6 + 8X_2X_7 + 11X_2X_8 + 13X_6X_7 + 16X_7X_8 \quad (3.27)$$

Table 3.7. Estimated ratio of self-compacting ash-containing concrete properties.

Concrete strength, MPa	Concrete class by frost resistance (F)	Concrete class by watertightness (W_t)
20	100 ... 150	4 ... 6
25	150 ... 200	6 ... 8
30	200 ... 250	8 ... 10
35	250 ... 300	10 ... 12

Analysis of model (Eq. 3.27) shows that self-compacting ash-containing concrete frost resistance largely depends on W/C and concrete mixture water content (Fig. 3.12). An increase in the values of the first and second factors as a result of higher capillary saturation of concrete leads to a decrease in concrete frost resistance.

Frost resistance of self-compacting ash-containing concrete substantially depends on the choice of the ash part in the mix of fine aggregates. For the selected experiment planning conditions the choice of ash part should be carried out considering all the investigated technological factors. In a common case, as follows from model (Eq. 3.27), to increase self-compacting concrete frost resistance, the ash part can be increased (Fig. 3.12). This situation can be explained by the fact that an increase in the ash part in the fine aggregate mix leads a higher ratio between the absolute volume of the binder (ash + cement) to that of voids in sand, as a result the thickness of the water shell surrounding the clinker grains decreases. Accordingly, the volume of pores in concrete and their size are reduced. It is suggested that this situation is true for concrete, in which ash is characterized by low carbon particles content, because increase in carbon particles content adversely affects the concrete frost resistance.

Based on analysis of graphical dependences (Fig. 3.12), obtained according to the frost resistance model (Eq. 3.33), it is evident that a decrease in W/C, S_a and M_f, the positive effect of increase in r_2 on frost resistance decreases. The increase in self-compacting concrete frost resistance due to the ash part growth is most noticeable at maximum values of these technological factors.

3.4 Proportioning and Optimization of Self-compacting Concrete Compositions

The method for calculating the self-compacting ash-containing concrete composition is based on the ash content constancy rule formulated above. It was previously shown that the optimal ash content can be set constant over a wide range of self-compacting concrete strength. The optimum ash content for normally hardened concrete is 150 ... 200 kg/m^3. More accurate ash content value can be found by taking into account the specific conditions of concrete production and application.

Joint solution of strength models (Eq. 3.9), water tightness (Eq. 3.25) and frost resistance (Eq. 3.27) enables it to establish an approximate ratio for the main design properties of concrete with optimal ash content (Table 3.7).

A further solution of the design and optimization problem lies in finding the composition of concrete with required strength, providing a set of design indicators. For this purpose, the necessary water-cement ratio (W/C) depending on the required design concrete strength class is found, taking into account the specific surface of used ash (S_a) and the sand fraction (r_s) in the mix of sand and crushed stone (Fig. 3.13). To obtain a nomogram for determining the water-

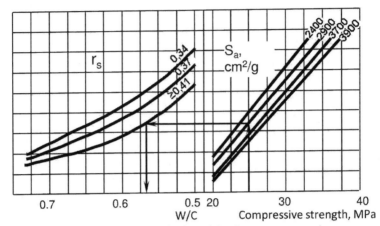

Figure 3.13. Nomogram for determining the water-cement ratio.

cement ratio, a mathematical model of concrete compressive strength at 28 days is used (Eq. 3.9).

To determine the value of the required factor r_s, cement efficiency coefficient (K_e) was chosen as optimization criterion. According to the model for K_e (Eq. 3.10), the optimal value of r_s was found depending on the selected ash specific surface (S_a) and content (D_a):

$$X_7 = 0.14X_4 + 0.9X_8 \qquad (3.28)$$

For go to natural values the following formula is used:

$$r_s = 0.41 + 0.07X_7 \qquad (3.29)$$

The minimum water demand of concrete mixture with SP admixture to achieve a self-compacting consistency is 180 kg/m³. However, based on the feasibility study, taking into account the specific conditions for self-compacting concrete production, the rational water content can be increased. Therefore, depending on the selected water demand, according to the given SP content mathematical model (Eq. 3.4), the necessary dry matter content of the superplasticizer (SP) that ensures the concrete mixture flow consistency is obtained (Fig. 3.14). Then the cement consumption, kg/m³ is found:

$$C = W/(W/C) \qquad (3.30)$$

and the total volume of sand and crushed stone in concrete (V_{s+cs}), l/m³ is calculated:

$$V_{s+cs} = 1000 - (C/\rho_c + W + A/\rho_a) \qquad (3.31)$$

where ρ_c and ρ_a are densities of cement and ash, respectively, kg/l.

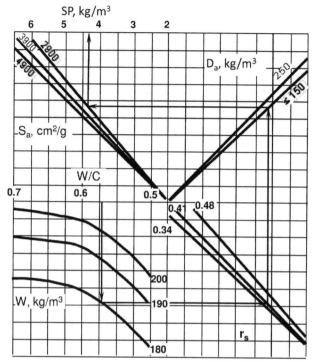

Figure 3.14. Nomogram for determining the content of SP.

Determine the sand content (S), kg/m^3.

$$S = V_{s+cs}\, r_s\, \rho_s \tag{3.32}$$

where r_s is the part of sand in the mix of sand and crushed stone;

ρ_s is the true density of sand, kg/l.

The crushed stone content (CS), kg/m^3:

$$CS = (V_{s+cs} - V_{s+cs}\, r_s)\, \rho_{cs} \tag{3.33}$$

where ρ_{cs} is the density of crushed stone, kg/l.

As an example lets design the composition of self-compacting ash-containing concrete with strength class C10/15, frost resistance F200 and water tightness grade W6 using Portland cement (32.5 class, $\rho_c = 3.1$ kg/l), quartz river sand ($\rho_s = 2.65$ kg/l), crushed granite stone fraction 5...20 mm (p = 2.65 kg/l), fly ash of the Ladyzhinstate district power station ($S_a = 2900$ cm^2/g; $\rho_a = 2.5$ kg/l) and SP superplasticizer.

The selected ash content is 150 kg/m^3, and the concrete mixture water demand is 180 kg/m^3. By Eq. (3.29) the optimal part of sand in natural value is

$$r_s = 0.41 + 0.07 \cdot 0 = 0.41$$

According to Table 3.7, a set of design requirements for concrete providing compressive strength of 25 MPa is found.

Following Fig. 3.12, the required water-cement ratio $W/C = 0.57$. The content of SP-4 is found using Fig. 3.14—it equals 4.46 kg/m^3. The cement consumption is obtained from Eq. (3.30):

$$C = 180/0.57 = 315 \text{ kg/m}^3$$

According to Eq. (3.31), the total volume of sand and crushed stone is:

$$V_{s+cs} = 1000 - (315/3.1 + 180 + 150/2.5) = 685.5 \ l/\text{m}^3$$

The content of sand following Eq. (3.32) is

$$S = 658.5 \cdot 0.41 \cdot 2.65 = 715 \text{ kg/m}^3$$

And the crushed stone content according to Eq. (3.33) is

$$CS = (658.5 - 658.5 \cdot 0.41) \, 2.65 = 1030 \text{ kg/m}^3$$

4

Metakaolin as Mineral Admixture for Cement-based Composites

Metakaolin, a product of kaolin clays calcination, a pozzolanic mineral admixture can serve as highly reactive cementitious component of mineral binders, concrete and mortars. The most applicable areas of metakaolin use in cement-based composites are self-leveling and self-compacting concrete, high-performance concrete; this is due to high aesthetical appearance of metakaolin—architectural concrete and finishing plasters.

4.1 General Information about Metakaolin as Mineral Admixture

4.1.1 Genesis and Properties of Kaolin

Kaolin is soft white clay. It is also called china clay. It is an essential ingredient in the manufacture of porcelain and is widely used in the making paper, rubber, paints, and many other products [17-19]. Kaolin clays are wide spread in many countries of the world. The largest reserves of kaolin clays are located in India, Czech Republic, China, Ukraine, Turkey [20]. The world's largest reserves of premium kaolin, which usually has minimum impurites, high britness and fineness, are located in United States and Brazil [21]. Global reserves of kaolin clay are shown in Fig. 4.1.

Kaolinite is a basic mineral of kaolin (Fig. 4.2). Kaolinite structure was described for the first time by Pauling (1930). According to him there was suggested that it consisted of a "sheet of Si–O tetrahedra arranged in a hexagonal network with a superposed sheet of Al–(O,OH) octahedra, the two together forming a layer of composition" [22]. It is a layered silicate mineral with chemical composition $Al_2Si_2O_5(OH)_4$ (Fig. 4.3).

There are two main types of kaolin: primary and secondary. Secondary kaolin is derived from primary. Primary kaolin has a high level of structure excellence, whereas the secondary kaolin has a low one [23]. The results of

Global reserves of kaolin clays

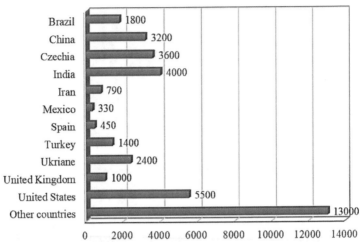

Figure 4.1. Global Reserves of Kaolin Clays, mln tons, following [18].

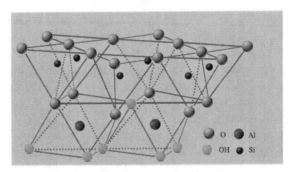

Figure 4.2. Kaolin Clay.

Figure 4.3. Kaolinite Lattice Structure.

researches prove the dependence of the pozzolanic activity of metakaolin on the structural regularity of kaolin clay [24]. Kaolinite is proved to have the highest potential to activation due to calcination rather than other clay minerals such as illite and montmorillonite [25, 26].

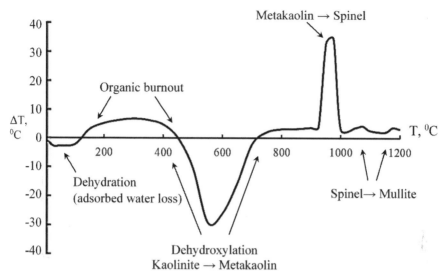

Figure 4.4. Diagram of Differential Thermal Analysis of Kaolinite.

Application of kaolin in many industries is mostly preceded by enrichment. Dry gravitational method is considered the most appropriate for further application of products of thermal treatment of kaolin [25].

Metakaolin is a dehydroxylated form of the clay mineral kaolinite, of which kaolin clays are rich in. To explain the pozzolanic effect and reactivity of metakaolin differential thermal analysis (DTA) of kaolinite can be considered (Fig. 4.4).

Figure 4.5. General View of High-reactivity Metakaolin.

The following equations illustrate the process under thermal treatment:

- formation of metakaolin [28]

$$Al_2O_3 \cdot 2SiO_2 \cdot 2H_2O \xrightarrow{\text{400–700°C}} Al_2O_3 \cdot 2SiO_2 + 2H_2O \qquad (4.1)$$

- formation of spinel

$$2(Al_2 O_3 \cdot 2SiO_2) \xrightarrow{\text{925–1050°C}} 2Al_2 O_3 \cdot 3SiO_2 + SiO_2 \ (amorphous) \quad (4.2)$$

- formation of mullite

$$3(2Al_2 O_3 \cdot 3SiO_2) \xrightarrow{\text{≥ 1050°C}} 2(3Al_2 O_3 \cdot 2SiO_2) + 5SiO_2 \ (amorphous)(4.3)$$

DTA curves for kaolinite have endothermic and exothermic effects [27]. The first endothermic effect reflects the evaporation of adsorbed water between 20 and 100°C with subsequent burnout of organic impurities up to 450°C. The second endothermic effect corresponds to the removal of crystalline (bound) water according to reaction and metakaolin formation between 450 and 600°C [29] (or 450–700°C according to [30]). Water removal is accompanied by destruction of the crystal lattice, amorphization and transition into metastable state. Endometric peak in this temperature range corresponds to dehydroxylation and appoximate water loss 13%.

Further heating is accompanied by the endothermal effect appeal, which is initially weak at 850°C and corresponds to formation of γ-alumina (γ-Al_2O_3), and subsequent heating to 900°C (911°C according to [8], 925–1050°C [28])—spinel formation of $2Al_2O_3 \cdot 3SiO_2$ and quartz: crystalline [24] or amorphous according to [28]). Exothermic effect at 1050°C corresponds to the formation of insoluble crystalline mullite $3Al_2O_3 \cdot 2SiO_2$ [28].

Metakaolinite partially decomposes, releasing amorphous γ-Al_2O_3 and SiO_2, before the 950°C [31]. The resulting phase is considered as spinel-type and is in fact associated with amorphous SiO_2 and γ-Al_2O_3 (crystalline) rather than with spinel.

4.1.2 Manufacturing and Properties of Metakaolin

The manufacturing process includes two main phases: thermal treatment within 450–700°C [30] and subsequent grinding (Fig. 4.5). The process of kaolin clay calcination to obtain metakaolin as a highly-reactive pozzolanic admixture is influenced significantly by heating temperature, heating duration and rate, cooling rate and ambient conditions [32, 33].

As a mineral admixture, metakaolin is considered as N type according to ASTM C618 [34] as if total amount of $SiO_2 + Al_2O_3 + Fe_2O_3$ is not less than 70%. The common name of the admixture is High-Reactive Metakaolin (HRM) to determine the pure admixture with high content of Al_2O_3 and SiO_2 (e.g., high kaolinite content) (Fig. 4.5) [35].

Complex activation during manufacturing metakaolin includes [25]:

- thermal activation (dehydroxylation of kaolinite and transition into metastable state) [36];
- mechanical activation (grinding metakaolin in grinding apparatus);

- chemical activation by $Ca(OH)_2$, Na_2SO_4, $CaSO_4$ (beta anhydrite) and NaOH)[37].

Pozzolanic activity of metakaolin varies on its purity [33]. This parameter is determined according to EN 196–5:2011 [38], as well as pozzolanic activity determined by the compressive strength of the specimens with and without a mineral admixture [39] or thermal analysis [40].

There is research that compares the pozzolanic activity of metakaolin to natural volcanic pozzolans and artificial pozzolans obtained at clay calcination [39,41]. Other research considers metakaolin as a mineral admixture with a high pozzolanic activity compatible to silica fume [42]. Table 4.1 gives comparative values of the pozzolanic activity of mineral admixtures determined by saturated lime tests, which gives the idea about the higher activity of the substance containing a higher content of active Al_2O_3, as it is able to form the products with a higher number of CaO molecules comparing to SiO_2 [25]. Similar research results are demonstrated by Sabir et al. [43].

Table 4.1. Pozzolanic Activity of Mineral Admixtures, mg of $Ca(OH)_2$ per 1 g of Admixture [25, 43, 44].

Mineral admixture	Calcined bauxite	Silica fume	Blast furnace granulated slag	Fly ash	Metakaolin
Pozzolanic activity	537	427	300	875	1050

The data given for laboratory and commercial types of metakaolin pozzolanic activity, determined by a direct modified Chapelle test, varies from 910 to 1560 mg $Ca(OH)_2/g$ [45]. Testing the pozzolanic activity by different methods proves that MK was a more pozzolanic material according to the Frattini Test and strength activity test (SAT), than fly ash and silica fume [46].

Resent research recognizes the problem of the sustainability of metakaolin as resources and energy consuming mineral admixture [47–50]. The sustainability issues can be solved in several aspects. The main one is in the reduction of cement binder due to the partial replacement for metakaolin. The reduction of environmental impact due to mining, thermal treatment and grinding can exclude enriching the clay and application of poor kaolin clays [47].

According to the available data, the total embodied energy of high-reactivity metakaolin and Global Warming Potential (GWP) are compatible to those for Portland cement (Table 4.2), whereas the embodied CO_2 is almost 3 times lower than that for Portland cement. The embodied energy values and embodied CO_2 for mineral admixtures made of industrial wastes are significantly lower due to the absence of the processes of mining, calcination and griding.

Companies that produce metakaolin, are mostly concentrated in Europe and USA. The main world companies are BASF (Metamax®), Imerys, I-Minerals, SCR-Sibelco, Thiele Kaolin, Advanced Cement Technologies (PowerPozz®), and others. Metakaolin production increased from 219000 MT in 2011 to 268000 MT in 2015, with an average growth rate of 4.47% [51].

Table 4.2. Environmental Impact of Portland Cement and Mineral Admixtures (based on aggregated data [48–50]).

Material	Embodied energy, MJ/kg [48]	Embodied energy, MJ/kg [49]	Global Warming Potential, CO_2 eq/kg [49]	Embodied CO_2, kg CO_2/ton [51]
Portland Cement	5.50	4.50	0.74	930
Metakaolin	-	5.99	0.034	330
Fly ash	0.10	-	-	4
Ground-granulated blast-furnace slag	1.60	-	-	52
Silica fume	0.036	-	-	14

The worldwide market for metakaolin is expected to grow at a compound annual growth rate of roughly 4.8% from 2019 to 2024. There is a prognosis of growth reaching 150 million US$ in 2024, from 110 million US$ in 2019, according to a Global Info Research study [51].

4.1.3 Application Areas of Metakaolin

One of the earliest known applications of metakaolin for construction was in the early 1960s for the dams in Brazil [52]. The role of metakaolin was in avoiding concrete bleeding and the improvement of concrete corrosion resistance [25, 27]. Metakaolin was also applied in 1960–70s for expanding hydraulic binders, consisted of Portland cement, metakaolin and dihydrate gypsum [53] and as a component of autoclaved slaked lime—metakaolin binder [25, 54]. Industrial manufacturing of metakaolin as an active mineral additive for mineral binders and concrete is comparatively recent: it started around in 1990s [44, 55].

Commercial metakaolin is usually considered as a component of architectural concrete, finishing plasters and component of restoration materials due to is high aesthetics characteristics. Metakaolin mostly finds application in cement-based mortars and concrete. In this area metakaolin is often compared to silica fume.

Table 4.3. Main areas of Metakaolin Application in Construction Materials Industry.

Application area	References
Alkali activated materials and geopolymers	56–60
Portland cement-based binders, concrete and mortars	25, 41–44, 53, 55, 61–68
Gypsum-based composites	69–71, 73
Lime-based composites	54, 71–74
Composites based on industrial wastes	75–79

However, metakaolin has some advantages over silica fume:

- as it is not an industrial waste, it provides higher uniformity and possibility in variation and optimization of its properties;

- there are many kaolin clays deposits in the world, forming a potential raw base for production of metakaolin;
- specific surface of metakaolin is significantly lower than that of silica fume, it causes lower water consumption and higher effect of water reducing admixtures;
- pozzolanic activity of metakaolin in cement systems is slightly lower or in some cases, higher than silica fume (see Table 4.1).

In our opinion one of the most applicable areas for the combination of superplasticizers and metakaolin as pozzolanic mineral admixtures is self-compacting and self-leveling high strength concrete, as if it permits it to reveal the potential of combined effect of chemical and mineral components on set of concrete properties.

4.1.4 *Multicomponent Admixtures Containing Metakaolin*

There is experience in producing multicomponent additives containing high-range water reducers (HRWR) or superplasticizer and mineral admixtures, like silica fume (SF) [80]. Along with many advantages there are a lot of problems with the manufacturing and application of composite admixtures based on SF:

1) absence of their own silica fume suppliers in many countries comparing to the wide potential raw base and available suppliers of metakaolin;
2) low manufacturability of silica fume due to high dispersity and corresponding low bulk density, and need in granulation to make it transportable;
3) high specific surface of silica fume causes significant water consumption growth of fresh concrete, with requires adding comparatively expensive high-range water reducers of polyacrylate (PA) or polycarboxylate (PC) type; as if melamine formaldehyde (MF) and naphthalene formaldehyde (NF) have lower efficiency for high-strength self-compacting concrete.

There are known methods of manufacturing granulated composite additive, based on silica fume and its combinations with fly ash, metakaolin and superplasticizer. Each granule is an agglomeration of ultrafine particles of mineral components, uniformly coated with a solidified adsorbed film of superplasticizer solution and other organic components [80]. The adsorption film "glues" the mineral admixtures particles together, forming granules, which are strong and stable in air-dry conditions. Being water-soluble, the film causes rapid disaggregation when mixing with water during concrete preparation.

The authors have studied a method of adding superplasticizer and metakaolin on properties of fresh and hardened concrete. Portland cement CEM I 42.5R, quartz sand with fineness modulus 1.5, metakaolin with a specific surface area 16700 cm^2/g and superplasticizer of NF type with defoaming agent were applied. Granite crushed stone with proportion 5–10 mm (30%) and 10–20 mm (70%)

Table 4.4. Compositions and Results of Concrete Testing Depending on the way of Introduction Admixtures.

#	Concrete composition, kg/m³									Slump, cm	τ^r_{2-18}, h	f_c, MPa, at the age, days		
	PC	SP (NF)	MK	Composite additive			S	CS	W			3	7	28
				total	SP	MTK								
1	500	6.0	44.0				612	1072	190	22	2.1	42.2	62.5	76.5
2	500			50	6,0	44.0	612	1072	190	23	2.4	48.9	66.5	83.3
3	500	7.2	52.8				610	1105	175	22	1.5	49.0	63.9	86.9
4	500			60	7.2	52.8	610	1105	175	23	1.9	48.3	72.5	92.7

Note. PC – Portland cement, SP (NF)-superplasticizer (naphthalene-formaldehyde type), MK – metakaolin; S – sand; CS – crushed stone, W – water.

was applied as coarse aggregate. Table 4.4 gives comparative characteristics of fresh and hardened concrete with separate additions of the admixtures (batches 1 and 3) and simultaneous adding as components of composite additive (batches 2 and 4).

As can be seen from Table 4.4, simultaneous adding of superplasticizer and metakaolin to the concrete composition as components of complex additive allows to increase the fresh concrete workability retention τ^r_{22-18} (time of slump reduction from 22 to 18 cm) and concrete strength. Similar data were obtained in the case of using the complex modifier based on silica fume [80]. This effect was explained by the gradual introduction of the composite additive into the fresh concrete. At the same time, adding SP directly into the mixture as water solution leads to its rapid adsorption on cement grains. Metakaolin particles retard the process of SP adsorption in composite admixture, it leads to increasing the workability retention [81]. Due to the higher homogeneity of the mixtures, the strength of concrete with a complex admixture is slightly higher.

4.2 Properties of Metakaolin and other Components of Cement Pastes and Self-leveling High-strength Concrete

4.2.1 Properties of Metakaolin as Mineral Admixture

At the first stage of the research four types of metakaolin have been applied. Three samples had a Ukrainian origin based on kaolin by AKW Ukrainian Kaolin Company (Glukhivtsy, Vinnitsa region, Ukraine). These samples were lettered as MK_1, MK_2, MK_3 and varied by chemical composition and physical properties (Tables 4.4–4.6). Another sample had been manufactured by the Engelhard Corporation (currently BASF SE, Iselin, New Jersey USA) as a trademark Metamax® EF [82], the sample was lettered as MK_4. Chemical

Table 4.5. Chemical Composition of Metakaolin Samples.

	Sample	MK_1	MK_2	MK_3	MK_4
Oxides content, %	SiO_2	52.5	54.6	52.5	53.0
	Al_2O_3	42.20	40.25	42.20	43.00
	Fe_2O_3	0.34	0.78	0.34	< 1.20
	TiO_2	0.70	1.08	0.70	< 1.5
	CaO	0.30	0.28	0.30	0.10
	MgO	0.25	0.28	0.25	< 0.1
	MnO	0.01	0.01	0.01	-
	Na_2O	0.10	0.08	0.10	< 0.05
	K_2O	0.90	0.60	0.90	< 0.4
	P_2O_5	-	0.04	-	-
	LOI	0.50	0.48	0.50	< 1.0

composition of metakaolin samples corresponds to ASTM 618 according to which content of $SiO_2+Al_2O_3+Fe_2O_3$ in pozzolan of N type is not less than 70% (see Table 4.5).

Kaolinite of Glukhivtsy deposits consists of tabular lamellar crystals, which mostly have half cut [25,83]. Results of scanning electronic microscopy (SEM)

Figure 4.6. Microstructure of Kaolin ×30000 (a), ×40000 (b) and Metakaolin ×400000 (c).

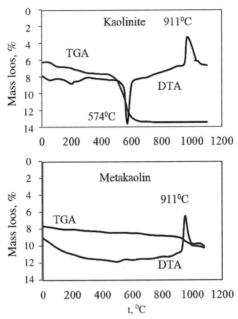

Figure 4.7. DTA and TGA curves of Kaolinite and Metakaolin (t = 700°C).

data confirms these research results (Fig. 4.6a,b). At the same time metakaolin consists of partly amorphized crystalline mass (Fig. 4.6c).

Results of TGA and DTA of kaolininte sample (Glukhivtsy deposit) (a) and metakaolin calcined at t = 700°C based on it (b) are shown in Fig. 4.7. DTA curves of kaolinite represent two endothermic and one exothermic effect. It meets the data given in subsection 4.1.1.

The mass loss of about 13% corresponds to the dehydration process that demonstrates an abrupt drop at TGA curve.

The absence of abrupt mass loss at TGA for metakaolin sample curve reflects the absence of chemically bound water in it. As well as an absence of endoeffects at DTA confirms complete dehydration of kaolinite, and exothermal effect at 911°C reveals the absence of the crystallization process and transition into a low-soluble state.

The amorphous state of metakaolin is evident from X-ray diffraction (XRD) data of kaolinite and metakaolin (MK$_1$) (Fig. 4.8).

Reflexe characteristic for minerals of kaolinite group corresponds to 0.714–0.720 nm and 0.357–0.360 nm (Fig. 4.8a). These reflexes either become less evident or disappear at thermal treatment up to 550–700°C (Fig. 4.8b,c). That proves the destruction of the crystalline lattice of kaolinite and transformation of the mineral into an amorphous state. It is confirmed by the other research data [84, 85]. Absence of diffraction peaks at d/n = 0.714; 0.357; 0.234; 0.149 nm for metakaolin X-ray curve (Fig. 4.8c) gives the idea about the transition into

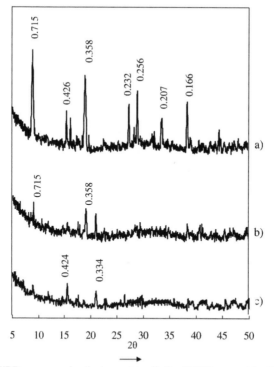

Figure 4.8. XRD curves: a – kaolin; b – metakaolin (t = 550°C); c – metakaolin (t = 700°C).

metastable state. The reflexes of metakaolin at 700°C (0.334; 0.424 nm) provide the evidence about insignificant amount of quartz.

Samples of thermally treated kaolin (metakaolin) were ground to receive dispersity, equivalent to sieve No.0045 residues less than 0.3%. The method of laser granulometry was used for the determination of the metakaolin's particle-size distribution of. Results are given in Table 4.6 and in Fig. 4.9.

Metakaolin samples have particle sizes up to 60 μm (Fig. 4.9 and Table 4.6). Metakaolin samples $MK_1 - MK_3$ have significant content of particles between 1 μm and 60 μm, whereas there is the lack of medium fraction (5–10 μm). MK_4 is more uniform with dominance the range of particle size 0.5–5 μm. Specific surface of metakaolin was calculated based on the particle size data. Pozzolanic activity of metakaolin was determined by the amount of free CaO, adsorbed

Table 4.6. Particle-size Distribution of Metakaolin Samples.

Sample	< 63 μm	< 30 μm	< 15 μm	< 5 μm	< 2 μm	d_{50}, μm	d_{90}, μm
MK_1	100	87.26	74.28	55.63	38.03	3.62	34.65
MK_2	100	98.49	88.20	61.60	43.10	2.83	16.30
MK_3	100	97.02	85.45	64.64	48.30	2.18	18.97
MK_4	100	100	99.36	92.20	73.00	1.04	4.30

Note. d_{50} and d_{90} are diameter of conventional cell, to pass 50% at 90% (vol.) of metakaolin sample correspondingly.

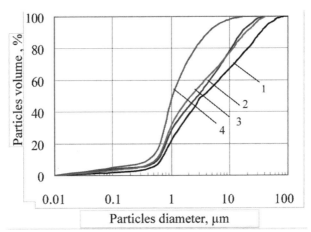

Figure 4.9. Particle Size of Metakaolin: $1 - MK_1$; $2 - MK_2$; $3 - MK_3$; $4 - MK_4$.

Table 4.7. Physical and Chemical Properties of Metakaolin.

Sample	Specific surface (by mass), m²/kg	Density, g/cm³	Bulk density, kg/m³	Normal consistency of metakaolin paste, %	Pozzolanic activity, mg/g (by CaO absorption)
MK_1	1380	2.50	400	43	19
MK_2	1670	2.50	410	46	26
MK_3	1800	2.50	350	54	25
MK_4	2590	2.50	280	60	42

by metakaolin. Physical and chemical properties of metakaolin are given in Table 4.7.

The higher specific surface of metakaolin is the higher is its water consumption and pozzolanic activity. The reason for these dependences is an opening the new active surfaces of metakaolin grains and possible amorphization during deep grinding.

4.2.2 Properties of other Materials Applied in Research

Portland cement. Locally produced Portland Cement PC CEM I 42.5, DIN 1146, EN-196 was used. ASTM Type I, normal Portland cement is usually applied for self-leveling and self-compacting high-strength concrete [90–91]. The chemical and mineral composition and properties of cement are given in Tables 4.8 and 4.9. The result of several research studies [86, 87] state, that application of highly-reactive mineral admixtures are mostly effective for pure clinker cements, as cements of other types contain mineral admixtures (slag, fly ash), which react to clinker minerals, forming additional hydrated compounds. That can lead to reduction of the metakaolin effect as an active mineral admixture. Portland Cement PC_1 is medium-alumina cement by its mineral composition and is considered typical

Table 4.8. Chemical composition of Portland cement, % by weight.

Sample	Chemical composition, %								
	CaO	SiO_2	Al_2O_3	Fe_2O_3	FeO	MgO	Cl^- ion	LOI	IR
PC_1	67.15	21.70	5.36	4.10	-	0.74	-	0.34	0.28
PC_2	65.58	22.05	4.73	4.42	0.06	1.62	0.012	0.28	0.10
Sample	Minerals composition, %								
	LSF		SR		AR	C_3S	C_2S	C_3A	C_4AF
PC_1	94%		2.34		1.31	65.17	13.06	7.26	12.46
PC_2	91%		2.41		1.07	60.01	18.30	5.03	13.43

Note. LOI – loss on ignition; IR – insoluble residue; LSF – lime saturation factor; SR – silica ratio; AR – alumina ratio

Table 4.9. Physical and Mechanical Properties of Portland Cement.

No.	Property	Units	Value	
			PC_1	PC_2
1	Fineness (No.008 sieve fraction residue), %	%	5	1
2	Specific surface by Blaine	cm²/g	3300	3500
3	Normal consistency	%	24	27.5
4	Setting time - initial - final	h-min h-min	$1-35$ $3-45$	$3-00$ $5-00$
5	Uniformity of volume change	-	Meets the requirements	Meets the requirements
6	Ultimate strength at 2/28 days: - flexural - compressive	MPa	5,65/8.75 23.70/54.0	4.83/9.70 19.38/53.80
7	Gypsum content (SO_3)	%	3.5	5.5

for the cement industry. Portland cement PC_2 is low-alumina cement (Table 4.8). Physical and mechanical properties of the cement are given in Table 4.9.

Higher values of early strength for PC_1 can be explained by the higher total content of C_3S and C_3A. Apart of it, PC_2, due to higher specific surface, has higher water consumption.

Aggregates applied for concrete manufacturing. Quartz sand (S) was used as fine aggregate (Table 4.10). There applied sand with fineness modulus was 1.6–1.8 (S_1) and 2.3–2.4 (S_2).

Granite crushed stone (CS) was used as coarse aggregate. It consisted of mix of fractions: size 5...10 mm (30...40%) and 10...20 mm (60...70%) to obtain minimum void ratio (voidage) (Table 4.10) [88]. According to particle size distribution data, both samples of sand meet the requirements to sieve residues (Fig. 4.10).

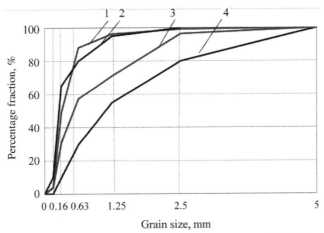

Figure 4.10. Particle size distribution of sand samples: 1 – sand S_1; 2 – lowest limit of fineness modulus (1.5); 3 – sand S_2; 4 – highest limit of fineness modulus (3.25).

As it is seen from Fig. 4.10, S_1 has a higher content of fine particles that increases water consumption. However, as some researches show, fine-grained sand with a percentage of particle size less than 0.14 mm up to 15% is suitable for self-leveling concrete from the point of low water bleeding [88, 89]. The

Table 4.10. Properties of Quartz Sand and Crushed Stone.

No.	Property	Units	Value		
			S_1	S_2	**CS**
1	Fineness modulus	-	1.6–1.8	2.3–2.4	-
2	Density (grain density)	g/cm³	2.64	2.65	2.74 (2.7)
3	Bulk density	kg/m³	1,440	1,530	1,510
4	Void ratio	%	45.5	42.3	44.1
5	Dusty and clayey particles content	%	≤ 1.5	≤ 2.5	≤ 0.5
6	Water consumption	%	8	6	2.5

fine fraction of sand is of great importance and the minimum particle size is recommended to limit with 75 μm in Japan and 125 μm in Europe [90].

Application of medium-grained and coarse-grained sand is recommended for production high-strength concrete [88]. Well graded combination should be applied for manufacturing high-performance concrete [91].

Superplasticizers. The following naphthalene formaldehyde (NF) type superplasticizers have been applied: S-3 (SP_1) and S-3 with defoaming agents (up to 0.01% by mass, SP_2). Defoaming agent permits to avoid excessive air-entraining in concrete with a high dosage of superplasticizer and metakaolin and also to increase the concrete strength as a result of additional densification of fresh concrete. The content of active substance in SP of NF type is higher than

69%, the ash content is up to 38%, pH of water solution at 2.5% concentration is 7–9.

Polycarboxylate (PC) type Melflux 2651F superplasticizer (SP_3) was also applied. Drying loss of SP_3 is up to 2%, the pH value of 20% solution is 6.5–8.5. Dosage recommendation for all the superplasticizers is up to 1.0%. All the superplasticizers are powder like substances and are F type admixtures-high range water reducers (Table 4.11) [92].

Table 4.11. Properties of Superplasticizers.

Superplasticizers	SP_1	SP_2	SP_3
Type of SP	NF	NF	PC
View	Powder, dark brown	Powder, brown	From yellowish to brownish
Bulk density, kg/m³	400–800	400–800	300–600
Recommended dosage, % by cement weight	0.1–1.0	0.1–1.0	0.05–1.0
Reduction of water consumption	Up to 20%	Up to 25%	Up to 40%

4.2.3 Effectiveness of Metakaolin and Composite Metakaolin-based Admixtures

Basic Approaches to Determining Effectiveness of Mineral Admixtures

There are different approaches for determining the effectiveness of mineral admixtures such as pozzolanic activity (see subsection 4.1.2), relative strength of concrete with admixture comparing to control one [39]. Due to high water consumption, most of the active mineral admixtures are effective only at the presence of high superplasticizers dosages, even if they partially substitute cements [55]. Therefore, strength increasing at raised SP dosage and reduced cement content are important factors displaying efficiency of admixture.

When superplasticizer and mineral admixture are combined, the important aspect of simultaneous application in cement-based systems is their compatibility with each other and with cement. Determining the thermokinetic parameters of cement hydration is an effective method for estimation of compatibility [61]. These parameters allow for predicting a series of properties for cement and concrete. However, such methods along with its high accuracy require special equipment.

There are other known parameters for measuring effectiveness of mineral admixtures as well as their combinations with HRWR [39, 47, 86, 94–95]. The parameters, representing the combined effect of cement and admixtures on flowability of fresh concrete and concrete strength, are the most adjustable to real manufacturing terms [86, 95, 96].

Deterministic dependencies for calculation of these parameters can be incomplete as they do not take into account some factors, whereas stochastic equations consider probability distribution of factors. Therefore, it was suggested to use a combination of deterministic equations and experimental models for estimation of constituent materials efficiency for concrete and mortars. A method for determining the efficiency of constituents was developed by Prof. V.G. Batrakov et al. [97]. This method was selected due to the simplicity of its implementation and unambiguity of the received results interpretation.

Methods of Testing

Cement mortars of the following proportion have been prepared:

1) control – without any admixtures at weight ratio cement to sand as 1:3;

2) basic – with composite admixture consisted of metakaolin and superplasticizer at weight ratio C (Cement): A (Admixture): S (Sand) = 0.9:0.1:3. SP content varied in the range of 0–1.5%, MK content – 0–15%.

Sand with fineness modulus of 2.0 (as a mixture of sand #1 and #2 in a ratio of 1:1, see Table 4.10) was applied. Water/cementitious material (cement+admixture) ratio of all basic mortars was equal to the control one. Batches have been prepared by electric mixer according to standard requirements adapted to such mortars with admixtures [98]. Error of batching by weight was less than 1%. Cement, modifier (where required) and 1/3 of sand were mixed for 1 min. In the dry mix water was added, and then mixed for 1 min. Then 2/3 of sand was added and mortar was mixed during 3 min. Water-cement ratio in the control mortar was sufficient for flow spread diameter of 113–114 mm. There were moulded cube specimens with an edge of 7.07cm. Flow spread diameter for each batch and compressive strength at 28 days of normal hardening were obtained.

According to the research method, activity indexes of composite admixture consisting of superplasticizer and metakaolin, were calculated as follows:

1) flowability activity index, %, is

$$I_f = \frac{D_b - D_c}{D_c} \times 100 \tag{4.1}$$

where D_b is spread diameter of the basic batch, mm;

D_c is spread diameter of the control batch, mm.

2) mortar strength activity index at 28 days is

$$I_s = \frac{f_b^c - f_c^c}{f_c^c} \times 100 \tag{4.2}$$

where f_b^c is compressive strength of the basic specimens, MPa;

f_c^c is compressive strength of the control specimens, MPa.

For each batch 2 samples were tested.

Results of Activity Indexes Estimation

At the 1st research stage three sets of experiments with different types of Portland cement and superplasticizer were conducted:

1st series: Portland cement PC_1, superplasticizer SP_1,

2nd series: Portland cement PC_2, superplasticizer SP_1;

3rd series: Portland cement PC_1, superplasticizer SP_2.

Each set included 4 batches, with specific samples of metakaolin used in each one at other constant conditions.

Control specimens had the following parameters:

when PC_1 used, W/C = 0.41; = 114 mm, f_c^c = 47.0 MPa;

when PC_2 used, W/C = 0.42; = 114 mm, f_c^c = 45.6 MPa.

Activity indexes of the specimens depending on the type of PC, MK and SP, are given in Table 4.12. The values of indexes confirm that the water consumption of metakaolin MK_4 is significantly higher than that for other samples due to its high specific surface (Table 4.12). Two opposite processes of increasing the pozzolanic activity of metakaolin and increasing water demand due to dispersity growth compensate each other (compositions 4, 8 and 12). Therefore, application of MK_4 can be effective from the viewpoint of flowability increasing at the combination with superplasticizer of higher range (e.g. PC type). At the same time the strength activities of specimens MK_3 and MK_4 are almost equal. Properties of mortar specimens with metakaolin samples MK_1-MK_3 were determined under further research.

Table 4.12. Activity Indexes of Mortar Specimens.

No. of series	No. of batch	PC	SP	MK	D_c, mm	D_b, мм	I_f, %	f_c^c, МПа	f_b^c, МПа	I_s, %
1	1			MK_1		188	65		50.5	8
	2	PC_1	SP_1	MK_2	114	171	50	47.0	54.4	16
	3			MK_3		156	37		55.4	18
	4			MK_4		105	-8		56.6	17
2	5			MK_1		183	61		48.8	7
	6	PC_2	SP_1	MK_2	114	165	45	45.6	51.8	14
	7			MK_3		150	31		52.4	15
	8			MK_4		100	−12		54.4	19
3	9			MK_1		197	73		51.2	9
	10	PC_1	SP_2	MK_2	114	185	62	47.0	55.4	18
	11			MK_3		172	51		56.0	19
	12			MK_4		117	3		56.1	19

Dependences of mortars activity indexes on the specific surface of metakaolin at $s \in [1{,}380; 1{,}800]$ m²/kg for each series of experiments can be approximated by parabolic equations. Depending on the type of Portland cement and superplasticizer used, the equations are as follows:

- for experimental series 1 (PC$_1$, SP$_1$):

$$I_f = -119(s \cdot 10^{-3})^2 + 313s \cdot 10^{-3} - 140 \tag{4.3}$$

$$I_s = -29(s \cdot 10^{-3})^2 + 116s \cdot 10^{-3} - 97 \tag{4.4}$$

- for experimental series 2 (PC$_2$, SP$_1$):

$$I_f = -126(s \cdot 10^{-3})^2 + 329s \cdot 10^{-3} - 155 \tag{4.5}$$

$$I_s = -39(s \cdot 10^{-3})^2 + 144s \cdot 10^{-3} - 117 \tag{4.6}$$

- for experimental series 3 (PC$_1$, SP$_2$):

$$I_f = -108(s \cdot 10^{-3})^2 + 290s \cdot 10^{-3} - 122 \tag{4.7}$$

$$I_s = -56(s \cdot 10^{-3})^2 + 201s \cdot 10^{-3} - 162 \tag{4.8}$$

where s is specific surface of metakaolin, m²/kg.

Figure 4.11 demonstrates graphical dependences of mortars activity indexes on specific surfaces of metakaolin (approximation curves 1 and 2). Comparing the values of activity indexes for mortars specimens of sets 1 and 2 (Fig. 4.11a,b) it should be mentioned that flowability index values are lower for PC$_2$ based mortars, and that is caused by the higher water demand of this type of cement.

The strength activity index of PC$_2$ based specimens is also slightly lower. That effect can be explained by lower content of C$_3$S and C$_3$A minerals, which have a higher hardening rate at an early age. Comparing the experimental results of sets 1 and 2, it can be seen that Portland cement PC$_1$ is more effective than PC$_2$, as it has a lower normal consistency. As it can be seen from comparing results of series 1 and 3, superplasticizers SP$_2$ is more effective. Substitution of SP$_1$ for SP$_2$ leads to increasing the mortars activity indexes under other constant conditions (Table 4.12, Fig. 4.11a, c).

Selection of Constituent Materials

For selection of optimal constituents, there should be established activity indexes of mineral admixture, that has a specific surface area compatible to metakaolin samples MK$_1$-MK$_3$. Such values had been known for the composite admixture consisted of silica fume and fly ash. Specified activity indexes are as follows: $I_f \geq = 60\%$ and $I_s \geq = 10\%$. These values were used for estimating the effectiveness of composite metakaolin-based admixture. 3 series of inequalities derived from Eqs. (4.3)–(4.4), (4.5)–(4.6) and (4.7)–(4.8) have been solved.

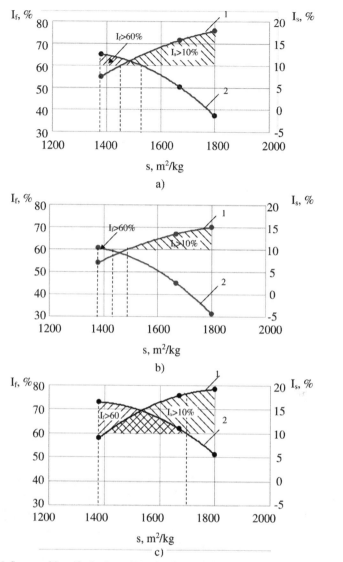

Figure 4.11. Influence of Specific Surface of Metakaolin on Activity Parameters of Composite Additive: a) PC_1, SP_1; b) PC_2, SP_1; c) PC_1, SP_2; 1– approximation curve of flowability activity index I_f, %; 2– approximation curve of strength activity index I_s, %,

Applying a graphic approach to the series of inequalities, solution regions of inequalities' sets were calculated and ranges of *s* were determined (Fig. 4.11, shaded areas). Thereby:

- for 1st series of experiments (PC_1, SP_1):

  for $I_f \geq 60\%$ s $\in$ [1380;1530] m²/kg; for $I_s \geq 10\%$ s $\in$ [1450;1800] m²/kg;

- for 2nd series of experiments (PC_2, SP_2):

for $I_f \geq 60\%$ s $\in$ [1380;1400]m^2/kg; for $I_s \geq 10\%$ s $\in$ [1460;1800] m^2/kg;
 • for 3rd series of experiments (PC$_2$, SP$_1$):
   for $I_f \geq 60\%$ s $\in$ [1380;1680] m^2/kg; for $I_s \geq 10\%$ s $\in$ [1410;1800] m^2/kg.

The results demonstrate that activity indexes are achieved within a range of s $\in$ [1450;1530] m^2/kg, at a ratio of MK$_1$ and MK$_2$ samples 1:1 for the 1st series of experiments (PC$_1$ and SP$_1$).

When PC$_2$ and SP$_1$ are applied, simultaneous providing the indexes values is impossible (Fig. 4.11b). The obtained data meets the results of known research, proving that the superplasticizer and mineral admixture have a higher activity at medium alumina cements with low normal consistency. Therefore, Portland cement PC$_2$was not considered relevant for further research.

As it is shown in Fig. 4.11c, when PC$_1$ and SP$_2$ are applied, providing both activity indexes values is possible with the range of s $\in$ [1410;1680] m^2/kg. Superplasticizer SP$_2$ has a higher water reducing and strength increasing effects due to additional mortar compaction. MK$_2$ is relevant to meet the requirements on specific surface (s = 1670 m^2/kg). SP$_2$ has a higher water reducing effect on mortar and increasing strength due to densifying mixture, caused by defoaming agent adding.

Regarding the results of tests and calculations, the following constituent materials have been selected for further research: Portland cement PC$_1$, superplasticizer SP$_2$ and metakaolin MK$_2$, as they enable us to provide the required set of activity indexes.

Air-entraining of mortars

Air-content makes a positive influence on concrete and mortar durability. According to available data, air content of 5–12% resulted in improved durability without causing serious reductions in compressive strength for cement-based mortars [99]. At the same time reduced air bubbles content seems to be improving the mixtures rheology [100]. In any case attention should be paid to air entraining, since certain properties of supplementary cementing materials used for self-compacting (SCC) and self-leveling concrete and HPC greatly influence the air-void system stability [101].

At the 2nd research stage a series of comparative experiments have been conducted to study the influence of superplasticizer type on mortars air-content. SP$_1$ and SP$_2$, Portland cement PC$_1$ and metakaolin MTK$_2$, selected at the 1st stage were used. Composite admixture content was 10% by binder weight. Superplasticizer content was fixed at the level of 5, 10 and 15% by composite admixture weight (0.5, 1.0 and 1.5% correspondingly by binder weight).

3 batches for SP$_1$ and SP$_2$ with dosages of superplasticizer given above were prepared. The air content was determined by gravimetric method. Air content of control mortar was $V_a = 14.1\%$.

Air content dependence V_a, % on SP dosage had been approximated by quadratic equations:

• when SP_1 is applied:

$$V_a = -2.4SP_1^2 + 8.2\ SP_1 + 7.1 \qquad (4.9)$$

• when SP_2 is applied:

$$V_a = -1.3SP_2^2 + 3.7SP_2 + 8.7 \qquad (4.10)$$

Index of air content change can be determined as follows:

$$K_a = \frac{V_{a,c} - V_{a,b}}{V_{a,c}} \cdot 100 \qquad (4.11)$$

where $V_{a,c}$, $V_{a,b}$ are air contents, % of control and basic mortars correspondingly.

Comparative dependences of mortars air content and air content change indexes are shown in Fig. 4.12. As it can be seen, the introduction of superplasticizer leads to a reduction of air content compared to the control mortar due to increasing the mortars flowability by 25% and 28% when a defoamer is used. Thus, the increasing of superplasticizer content dramatically reduces its effect. At the point of increasing the dosage of SP_1 from 0.5 to 1.5% the mortar air content is almost equal to that of the control mortar regardless of high increasing of flowability. That confirms the known data about intensive air-entraining of HRWR at high dosages [86, 102]. It is expected that application of defoaming agent allows the obtaining of a slight decrease in air content even when high dosages of SP_2 are used and provides a higher plasticizing effect and higher strength of mortars and concrete.

The adequacy of the obtained equations was checked by calculating the adequacy dispersion, the design value of Fisher's criterion (F_d – criterion) and comparing the last with a theoretical one F_t. The regression equation is adequate for the given probability level if $F_d < F_t$. For the equations given above Fisher's criterion $F_p = 1.0 < F_T = 4.28$.

Therefore, Portland cement PC_1, superplasticizer SP_2 and metakaolin MK_2 were selected for further research, as they provided the required activity indexes values.

4.3 Hydration and Structure Formation of Cement Pastes with Metakaolin-based Admixture

Rheology and mechanical properties of cement matrix make a significant impact on the properties of concrete. Normal consistency, water retention, setting time, water absorption were tested according to standard methods [103]. Other relevant methods of testing are described admixtures below.

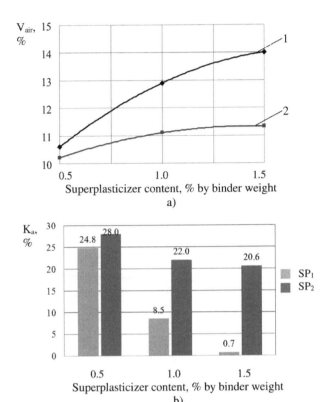

Figure 4.12. Influence of Superplasticizer type (NF) and its Content on: a) mortarsair content; b) air content index change 1- SP_1; 2- SP_2.

4.3.1 Normal Consistency and Water Retention of Cement Pastes

Water Retention of Cement Pastes

Normal consistency (NC) reflects the "tightly bounded structure of cement gel with dense spatial package of solid phase particles, coated by water" [104]. Normal consistency is an important indicator of the cement pastes physical characteristics. It has a linear dependence on ultimate water retention and depends on cement composition and dispersion, as well as the admixtures type and dosage.

To determine the influence of the metakaolin-based admixture on the normal consistency and water retention of cement pastes the research with appying selected components Portland cement PC_1, superplasticizer SP_2 and metakaolin MK_2 (see subsection 4.2.2) has been conducted. MK and SP contents were set as percentage of binder mass (Portland cement+metakaolin). Experiments have been replicated for each composition. Compositions of paste and mean values of normal consistency are presented in Table 4.13.

Experimental data was approximated by the following equation:

$$NC = (A \cdot X_1^2 - B \cdot X_1) + (NC^0 + C \cdot X_2/100) \qquad (4.12)$$

where *A, B* are coefficients, depending on the superplasticizer type; for SP: $A = 0.011$; $B = 0.053$;

X_1 is SP content, % by binder weight;

NC^0 is normal consistency of control cement paste (without admixtures), $NC^0 = 0.24$;

X_2 is metakaolin content, % by binder weight;

C is a coefficient, $C = NC_{MK} - NC^0$, $NC_{MK} = 0.46$, $C = 0.46 - 0.24 = 0.22$.

Table 4.13. Cement Pastes Compositions and Normal Consistency Values.

No.	SP, % by binder weight	MK,% by binder weight	Normal consistency (NC)	
			experimental value	calculated value
1	0	0	0.240	0.240
2	0.5		0.216	0.216
3	1.0		0.197	0.197
4	1.5		0.185	0.184
5	0	5	0.252	0.251
6	0.5		0.227	0.227
7	1.0		0.210	0.208
8	1.5		0.197	0.195
9	0	10	0.262	0.262
10	0.5		0.239	0.238
11	1.0		0.223	0.219
12	1.5		0.206	0.206
13	0	15	0.272	0.273
14	0.5		0.251	0.249
15	1.0		0.228	0.230
16	1.5		0.215	0.217

Equation (4.12) is adequate at 95% confidence probability: $F_d = 2.03 < F_t = 2.46$.

Figure 4.13 demonstrates the influence of MK and SP content on normal consistency. As it can be seen from Eq. (4.12), the dependence of NC on SP content can be approximated by a parabola, which means that SP effectiveness decreases as its content grows. Effect of metakaolin is linear and is independent of SP content. The minimum amount of water required for the formation of coagulation structures in cement paste, is [117]:

$$NC^{min} = 0.876 \, NC \qquad (4.13)$$

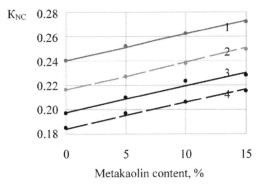

Figure 4.13. Impact of Metakaolin Dosage on the Normal Consistency of Cement Paste at SP Content:
1) 0%; 2) 0.5%; 3) 1.0%; 4) 1.5%.

When the water content is equal to NC, water is contained mainly in solvate shells. As follows from Table 4.13 and Fig. 4.13, when SP is added, NC^0 can be less than NC^{min}. In this case the effect of dilution, caused by the influence of SP, leads to a significant reduction in the solvate shells thickness and the releasing additional volume of free water. NC of cement paste with SP is close to the theoretical W/C, required for hydration, which is 0.21–0.23 [118]. Introduction of MK, which has a higher normal consistency than cement (see Table 4.13), leads to the subsequent growth of the cement paste water demand and decreasing or total leveling of SP water-reducing effect. However, as shown in Fig. 4.15, introduction of high dosages of SP (1–1.5%) leads to appreciable water content reduction of pastes containing metakaolin, compared to cement paste without admixtures. Replacing 5% of PC by MK at superplasticizer dosage of 1.5% leads to decrease of 20% in normal consistency; when 15% of PC is replaced by MK normal consistency reduction is 10%.

Water Retention of Cement Pastes

Water-binder ratio (W/B), which corresponds to the ultimate water retaining capacity of cement pastes, varies from 1.60·NC to 1.65·NC [104]. However, taking into account some possible vibrations of transportation it is considered from 1.30·NC to 1.35·NC [4,104].

It was determined the influence of MK and SP on water-retaining capacity of cement suspensions. Water bleeding coefficient (K_b, %) of cement suspensions with the SP and MK at water-binder ratio (W/B) equal to 1 was tested. K_b represents the volume of water separated:

$$K_b = \frac{a-b}{a} \cdot 100 \qquad (4.14)$$

where a is the initial volume of cement suspension cm³;

b is the volume of cement in settled suspension, cm³;

The water-retaining capacity (WR) of cement suspensions is as follows

$$WR = 1 - K_b \qquad (4.15)$$

WR, divided by 100, is W/B of settled suspension. Mean values of bleeding coefficients and water-retaining capacity are shown in Table 4.14.

Table 4.14. Water Bleeding Coefficient and Water Retaining Capacity of Cement Suspensions.

No.	SP, %	MTK, %	Water bleeding coefficient, % Water retaining capacity, % after, hours								
			0	0.5	1	1.5	2	2.5	3	3.5	4
1	0	0	0 100	25.4 74.6	36.8 63.2	41.9 58.1	44.2 55.8	45.2 54.8	45.6 54.4	45.8 54.2	45.9 54.1
2	0	10	0 100	13.2 86.8	19.0 81.0	22.0 78.0	23.9 76.1	25.5 74.5	25.9 74.1	25.9 74.1	-
3	0.5	10	0 100	17.6 82.4	25.5 74.5	29.0 71.0	30.0 70.0	31.1 68.9	32.0 68.0	32.4 67.6	32.5 67.5
4	1	10	0 100	21.0 79.0	30.0 70.0	35,2 64,8	37,8 62,2	38,0 62,0	38,7 61,3	39.4 60.6	39.5 60.5
5	1.5	10	0 100	23.5 76.5	34.5 65.5	41,6 58,4	43,8 56,2	44,0 56,0	45,5 54,5	45.6 54,4	45,7 54,3
6	1	0	0 100	30.0 70.0	45.7 54.3	54.5 45,5	56.0 44,0	57.5 42,5	58.9 41,1	58.9 41.1	-
7	1	5	0 100	25.0 75.0	38.5 61.5	43.0 57.0	47.0 53.0	48.5 51.5	48.8 51,2	48.9 51.1	49.0 51.0
8	1	10	0 100	19.0 81.0	30.0 70.0	34.5 65.5	38.0 62.0	38.3 61.7	38.9 61.1	39.0 61.0	-
9	1	15	0 100	14.5 85.5	24.0 76.0	25.5 74.5	27.0 73.0	28.5 71.5	28.8 71.2	28.9 71.1	-

The influence of water retention on the time of settlement can be described by hyperbolic dependence (Fig. 4.14) with horizontal asymptote as follows:

$$WR = 100 - (A + BX_1 + CX_2)(1 - 1/D^r) \qquad (4.16)$$

where X_1 is the SP content, % of binder weight;

X_2 is the MK content, % of binder weight;

τ is the sedimentation time, h.

A, B, C and D are experimental coefficients, which depend on properties of the applied materials (MK, SP); in the current research $A = 46;$ $B = 13;$ $C = -2;$ $D = 5$.

Equation (4.16) is adequate at 95% confidence probability: $F_d = 1.35$ < $F_t = 1.36$.

Figure 4.14 shows dependencies of water retaining capacity on the SP and MK content. With increasing SP content ordinate asymptote decreases (Fig. 4.14a); as MK dosage grows - ordinate asymptote increases too (Fig. 4.14b).

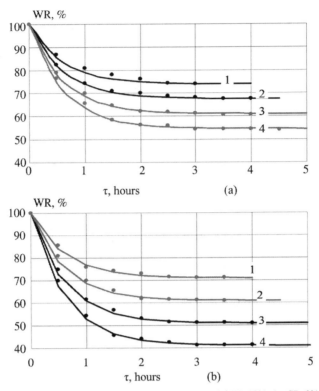

Figure 4.14. Kinetics of Cement Suspensions Water Retaining: a) MK=10%: 1 – SP=0%, 2 – SP=0.5%, 3 – SP=1%, 4 – SP=1.5%; b) SP=1%: 1 – MK=15%, 2 – MK=10%, 3 – MK=5%, 4 – MK=0.

The second part of Eq. (4.16) describes the water-retaining capacity dependence on time. As the suspension settlement time grows, water retaining capacity reduces close to asymptote. The presence of MK reduces suspension bleeding due to higher stability against sedimentation of smaller MK particles in comparison to Portland cement.

According to Table 4.14, the use of SP and MK admixtures allows us to obtain suspensions with WR = (2.5–3)·NC at the end of experiment. At the same time, the control suspension without admixtures at the end of the experiment has $WR_{lim} \approx 2 \cdot NC$, that value is close to the theoretical limit of cement water-retaining capacity (upper limit of thixotropy) [106]. That is caused by the higher water consumption of MK comparing to PC.

The water-retaining capacity of suspensions, made of binders with equal water consumption, has been compared. It was assigned that NC = 0.24 (as for control composition). Equal water consumption was achieved by varying MK content from 5 to 15%, SP content was calculated according to Eq. (4.12). Water retention was determined using Eq. (4.15). The results are shown in Fig. 4.15.

As it can be seen, the increasing of MK content at substantial SP dosage increase leads to the significant growth of water retaining capacity. At replacing

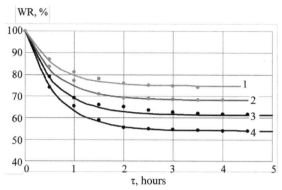

Figure 4.15. Kinetics of Water Retaining of Cement Suspensions with Equal Normal Consistency 24% calculated by Eq. (4.12): 1 – SP=0.72%, MK=15%; 2 – SP=0.45%, MK=10%; 3 – SP=0.21%, MK=5%; 4 – SP=0%, MK=0%.

5% of PC by MK WR = 61.3% (W/B = 0.613), at 15% — WR = 74.7% (W/B = 0.747). Due to possible vibration impact at adding MK the limit value of $WR_{lim} = 2NC$.

However, the testing of cement pastes at W/B = NC and W/B = 1 is not quite as informative, as if for self-leveling concrete NC < W/B < 1. Normally for high-strength concrete W/C < = 0.4 [91]. Therefore, we considered the properties of cement pastes at «true» $(W/C)_t$ (W/B), at which fresh concrete will have similar flowability to cement paste [6]. «True» W/C of concrete is calculated as

$$(W/C)_t = (W - W_{f.a} \cdot FA - W_{c.a} \cdot CA)/C \qquad (4.17)$$

where $W_{f.a}$ and $W_{c.a}$ are water consumption of fine aggregate and coarse aggregate correspondingly,

FA is fine aggregate content, kg/m³;

CA is coarse aggregate content, kg/m³;

W is water consumption of fresh concrete, kg/m³;

C is cement content, kg/m³

In the case of high-strength concrete W/C is normally lower than 0.4. Taking into consideration water consumption of fine and coarse aggregate for their moistening, actual W/C of cement paste in concrete is within the range 0.25–0.35 [6]. Considering ultimate water retention of control cement paste, the water-binder ratio selected for further research was

$$W/B = 1.35NC^0 = 1.35 \cdot 0.24 = 0.33. \qquad (4.18)$$

Flowability of Cement Pastes

The flowability of cement pastes was determined by the Vicat apparatus cone, which is compartible to Abrams cone for fresh concrete [107]. Each experiment has been conducted twice. Mean values of the flow spread diameter of cement pastes are shown in Table 4.15.

As in the case of normal consistency, dependencies of flow spread diameter are approximated by following regression equation:

$$D_f = (AX_1^2 + BX_1) - (CX_2 - D) \qquad (4.19)$$

where *A, B* are coefficients depending on the superplasticizer type; for the given SP A = –26.2; B = 60.5;

C, D are coefficients, depending on the type of metakaolin, its water consumption; for given MK C = –1.1; D = –3.4;

X_1 is the SP dosage, % by binder weight;

X_2 is the MK dosage, % by binder weight.

Equation (4.19) is adequate at 95% confidence probability: F_d = 3.08 < F_t = 3.51. Graphical dependencies of flow spread diameter of cement paste on SP and MK content are given in Fig. 4.16.

As it can be seen from Eq. (4.19) and Fig. 4.16, the dependence of flow spread diameter on SP content is parabolic. Increasing the MK portion leads to

Table 4.15. Influence of Metakaolin-based Admixture on Flow Spread Diameter of Cement Pastes.

No.	SP,%	MK,%	Flow spread diameter, cm	
			experimental value	calculated value
1	0.25	0	17.5	16.8
2	0.25	5	11.0	11.4
3	0.5		21.5	21.4
4	0.5	10	15.7	15.9
5	0.75		10.5	10.5
6	1.0		23.5	22.6
7	0.5	15	16.8	17.1
8	0.75		25.1	25.3
9	1.0		21.0	20.3

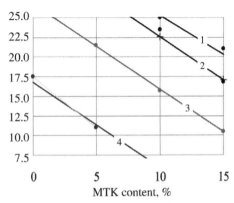

Figure 4.16. Influence of Metakaolin Content on the Flow Spread Diameter of Cement Pastes at SP Content:1) 1.0%; 2) 0.75%; 3) 0.5%; 4) 0.25%.

subsequent reduction of pastes flowability according to linear dependence. At a given value of water-binder-ratio (W/B = 0.33) and considered SP and MK dosages, flow spread diameter was 20 cm as the paste in this case is self-leveling.

Therefore, further research has been conducted at pastes with W/B = 0.33 and D_f= 20 cm at partial substitution of cement by 5, 10 and 15% for metakaolin. The required SP dosage to achieve the assigned flowability was determined according to Eq. (4.19). The condition $W/B < (WR)_{lim}$ is provided. Test results are given in Table 4.16.

Table 4.16. Cement Compositions with Metakaolin Based Admixture of Equal Flowability.

No.	W/B	D_f, cm	MK, %	SP, %	NC, (according to Eq. 3.1)	$(WR)_{lim}$
1	0.33	-	0	0	0.240	0.325
2		20	0	0.32	0.224	0.448
3			5	0.46	0.229	0.458
4			10	0.64	0.232	0.464
5			15	0.95	0.233	0.466

All the batches containing metakaolin and superplasticizer, meet the water retention ability requirement. It is confirmed by the data of cement paste water retention given in Table 4.16. Water retention was determined in % (Fig. 4.17). As it can be seen from the diagram, at increasing the metakaolin dosage in binder, water retention capacity grows at equal flowability. It can be concluded that metakaolin-based admixture permits us to obtain cement pastes without water bleeding even at raised water consumption comparting to control composition. The water retention capacity limit for pastes with metakaolin-based admixtures is achieved at about 2.0·NC.

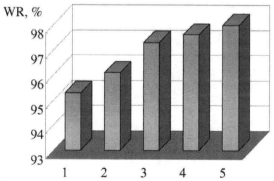

Figure 4.17. Water retention capacity of cement pastes: 1 – control; 2 – SP=0.32%, MK=0%; 3 –SP=0.46%, MK =5%; 4 – SP=0.64%, MK=10%;5 – SP=0.95%, MK=15%.

4.3.2 *Initial Structure forming of Cement Pastes*

General Information about Initial Structure forming Research Approaches

Initial structure forming process had been studied by different methods. The kinetics of initial structure forming of cement pastes was studied by the Vicat needle test and measuring supersonic waves at passage speed, resistivity ρ (conductivity σ) during 24 h starting with mixing time.

Standard setting time test of cement paste permits to fix initial and final setting time, e.g., conditional time of hardening process beginning, at that time the sensibility of the Vicat apparatus is comparatively low and the test duration is limited by the final setting time. The methods of plastic strength, specific conductivity σ (specific resistivity ρ), and supersonic waves velocity C_1 permit us to observe the structure forming both during the setting period as well as in the initial period of hardening.

According to current ideas, three main stages in the process of structure forming of the cement [108] are:

a) induction period (stage) following directly after mixing with water;

b) stage of initial structure forming or coagulation, i.e., setting period;

c) crystallization stage, i.e., hardening period.

Such division is conditional and it is reasonable to consider the prevalence of one process over another in experimental results discussion.

According to [109], limits of these stages are determined by distinctive time τ_1, τ_2, τ_3 – the points of qualitative changes in structure of cement system, shown in Fig. 4.18. Time τ_1 reflects the conditional final time of induction period and start of setting period, finished at τ_2. At that time aluminate structural

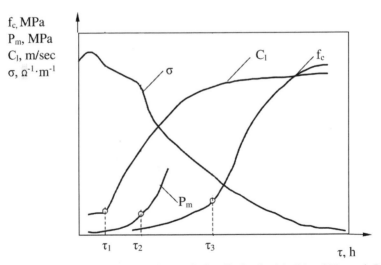

Figure 4.18. Kinetics of Strength Properties (P_m, f_c), Specific Conductivity (σ) and Ultrasonic Pulse Velocity (C_1) at Hardening of Cement Paste (based on [108]).

grid is predominantly formed. Crystallization growth of silicates as basic carriers of strength starts later and can be characterized by time τ_3.

Research Methods and Comparative Results

Plastic strength of cement pastes was determined by conic plastometer. The penetration depths was fixed by the indicator with a value of 0.01 mm at time range of 30 min. The P_m value was determined according to the following equation:

$$P_m = K \cdot \frac{P}{h^2} \qquad (4.20)$$

where K is a constant for the device, at cone angle of $\varphi = 30°$K $= 0.96$;

P is the cone weight with tightening weights, N;

h is the cone penetration depth, m, h $= 0.005$ m.

Specific conductivity of specimens (σ, $\Omega^{-1} \cdot m^{-1}$) was calculated according to the following formula:

$$\sigma = \frac{1}{\rho} = \frac{1}{R} \cdot \frac{l}{s} \qquad (4.21)$$

where ρ is specific resistivity, $\Omega \cdot m$;

R is resistance of the section under investigation Ω;

l is the basic length of the section, m;

S is cross-sectional area, m^2.

Ultrasonic waves velocity was determined by acoustic (impulse) method at frequency of 60 Hz (accuracy ± 0.8–1%), i.e., 0.25–0.5 mcsec for cement paste cubic specimens with 10 cm edge, hardened at normal conditions. The ultrasonic waves' velocity C_l, m/sec was determined as

$$C_l = l/t \qquad (4.22)$$

where l is the base of sounding, m;

t is the ultrasonic waves transmitting time, sec.

The test has been conducted by the method of through sounding with an interval of 30 min.

For results correlation the cement pastes parameters were compared by different methods (see Table 4.17). There results were obtained for normal consistency pastes at W/B $= 0.33$ (see subsection 4.3.1).

Available data on the influence of superplasticizer and mineral admixtures (silica fume and metakaolin) demonstrates that these admixtures have an insignificant impact on setting time compared to control compositions [81, 110]. The experimental data confirms the acceleration of the structure forming processes in cement pastes at SP presence and high C_3A content in clinker [111]. This phenomenon is explained by intensive adsorption of superplasticizer

Table 4.17. Comparison of Setting Time of Cement Pastes Measured by Vicat Apparatus, Plastic Strength Method and Acoustic Method.

No.	SP, %	MK, %	Setting time, h:min W/B = NC		Setting time, h:min W/B = 0.33		τ_1, ultrasonic pulse velocity test, h:min	τ_2, plastometric test, h:min
			Initial	Final	Initial	Final		
1	0	0	1:35	3:45	4:25	6:45	4:00	7:30
2	0.32	0	3:15	4:35	6:30	7:50	6:10	8:20
3	0.46	5	1:25	4:20	4:40	8:00	4:30	8:25
4	0.64	10	1:45	4:35	4:50	8:25	4:25	9:20
5	0.95	15	2:15	4:55	5:00	8:40	4:50	9:55

solution by aluminate compounds, and vice versa, for low-alumina cement SP which was not adsorbed intensively.

As low-alumina cement was applied in the experiments, it is observed that SP retards cement paste setting due to lower thickness of the solvate shells compared to high-alumina cements and releasing additional mixing water. The water can adsorb both at the grains of nonhydrated cement and hydrated new formations. Adsorption results in increasing of diffusive resistance and inhibition of reactions cement grains [86, 111]. At simultaneous introduction of MK and SP, terms of hydration are considered to be "tight" [112]. There is a close contact between the solid particles due to the increasing their concentration and higher specific surface area of MK comparing to Portland cement. Because of the physical and chemical water bounding, the thickness of water interlayers reduces. It leads to the acceleration of initial setting time. The intensity of new formations growth reduces with time, and the final setting time of pastes No. 3–5 is close to that of paste No. 2. As MK content increases, the paste structure forming processes retards due to adsorption of SP on particles of MK and blocking cement hydration. Similar phenomena are observed for pastes with W/B = 0.33. However, because of their higher water content, initial structure forming processes retard at 3–5 h (Table 4.17).

Plastic Strength

Figure 4.19 demonstrates most distinctive graphical dependencies of plastic strength change of the cement pastes with W/B = 0.33. According to [113], first plot sections up to τ_2 (white points) reflects the dominance of inverse thixotropic structure, which represents the spatial grid of irregular shape made of colloid particles, hydration products and clinker grains. P_m value slightly increases in this section.

From the other side, adsorption films reduce the adhesion forces between the particles, weakening the coagulation contacts. Therefore, pastes modified with SP and MK+SP, are characterized by a longer period of coagulation structure formation.

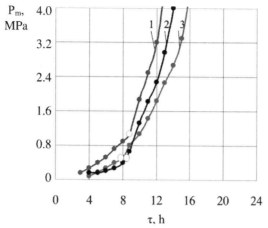

Figure 4.19. Kinetics of Cement Pastes Plastic Strength: 1 – SP=0.64%; MK=10%; 2 – SP=0.32%; 3 – control (without admixtures).

The second section reflects the period of the coagulation structure strengthening and beginning of crystallization. Metakaolin particles, coated by water adsorption films, fill the voids between the coarse cement particles, forming a viscoplastic medium. This medium has increased plastic strength in contrast to compositions 1 and 2. It should be noted that the time τ_2, determined by graphical dependences, is slightly longer than the final setting time, determined by the Vicat apparatus (Table 4.17). Subsequent structure-forming processes are characterized by intensive strengthening and the formation of new crystalline concretions (composition 3). The adsorption layer of SP is permeable to water, and its peptization effect increases the contact surface of cement and metakaolin with water. Therefore, after the initial retardation, the hydration processes are accelerated. The number of hydrated reaction products and concretion contacts between them increase to such extent that the system acquires higher structural strength. Cement hydration at a presence of reactive mineral admixtures and SP occurs at relatively low concentrations of Ca^{2+} [70]. Under such conditions, a large amount of highly dispersive low-base hydrosilicates appears in the hydrated cement medium. Therefore, it can be assumed that such an increase in plastic strength for paste 3 is caused by intensive binding of free lime by metakaolin particles to low-base calcium silicates.

The cone plastometer can be used as long as $P_m \leq 8$ MPa ($f_c \leq 1$MPa). Further study of pastes structure forming should be conducted to determine the samples compressive strength [109].

Compressive Strength

Compressive strength results of paste cube specimens with edge of 2 cm are shown in Fig. 4.20. Graphic dependencies indicate a more intensive crystallization of silicate reaction products at third stage of structure forming (white points indicate the time τ_3). It can be assumed that this phenomenon is also caused by

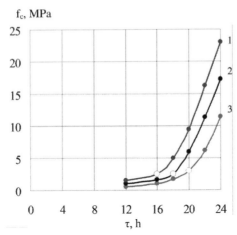

Figure 4.20. Compressive Strength of Hardened Cement Pastes: 1 – SP=0.64%; MK=10%; 2 – SP=0.32%; 3 – control (without admixtures).

the reaction of metakaolin with free lime and the formation of additional low-base stable calcium hydrosilicates.

Specific Conductivity and Resistivity

Kinetics of pastes specific conductivity, a quantity inverse to resistivity, also correlates to growth in plastic strength and compressive strength (Fig. 4.21a). Cement particles when interacting with water form an ionic medium capable of conducting an electric current; the electrical conductivity of such a medium increases in proportion to the degree of electrolytes dissociation (cement minerals in water) [34].

After measuring the change in electrical conductivity over time, a number of characteristic curves have been obtained. At the initial period conductivity increased and after a certain time, which is 0.5–1 h on average, it reaches its maximum, then the conductivity value changes insignificantly over time, and then it decreases (Fig. 4.21 a).

During the first 40 min, the process of hydrolysis of Portland cement clinker minerals (mainly C_3A and C_4AF) and formation of a saturated calcium hydroxide solution, increasing the conductive ions concentration per unit time occur [34, 120]. Under such processes conductivity of cement pastes increases. The maximum conductivity reflects the beginning of induction period [120]. During this period intensive water absorption by cement minerals, formation of the coagulation structure of cement pastes and crystal nuclei occur. Due to a smaller space between cement particles, the coagulation process in the control cement paste runs faster, which leads to shortening the induction period. In case of presence of the SP solution, adsorption films on the surface of cement particles, or cement and metakaolin particles, the hydrolysis processes are somewhat slowed down. The second decline of the curves reflects the end of the induction period and beginning of the structure formation, which is approximately characterized by the Vicat apparatus as an initial setting time. At later period, conductivity decreases monotonically until it reaches a minimum in the solidified mass.

Further reduction of conductivity to its minimum value is explained by minerals hydrolysis process (mainly C_3S). Pozzolanic activity of metakaolin causes more intensive growth of electrical resistance due to rapid binding of conductive ions (Fig. 4.21b). The nature of the conductivity curves correlates with the character of conductivity kinetics curves for the 'Ca(OH)$_2$–metakaolin' system given in [114].

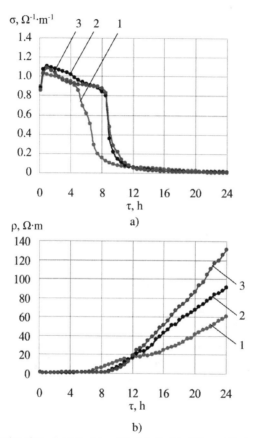

Figure 4.21. Kinetics of Conductivity (a) and Resistivity (b) of Cement Pastes: 1 – SP=0.64%, MTK=10%; 2 – SP=0.32%, MTK=0%; 3 – SP=0%, MTK=0%.

Ultrasonic Pulse Velocity

The ultrasonic velocity can determine the end time of the inductive stage in structure formation after mixing with water [115], which corresponds to the initial setting time. The method of through passage of longitudinal ultrasonic pulse was used during 24 h with measuring each 30 min. Results of the test are given in Table 4.17 and Fig. 4.22 (for W/B = 0.33).

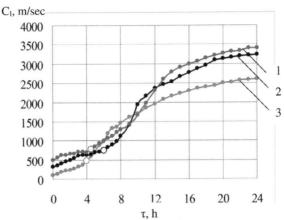

Figure 4.22. Kinetics of Ultrasonic Pulse Velocity at Initial Structure Forming: 1 – SP=0.64%, MTK=10%; 2 – SP=0.32%, MTK=0%; 3 – SP=0%, MTK=0%.

Curves of ultrasonic pulse velocity have sections, which correspond to the stages of structure forming. At the initial stage (the induction period) C_1 slightly increases (marked as horizontal section). Subsequent formation of the low-strength crystalline lattice of aluminate components significantly affects C_1: $dC/d\tau$ is maximum at this stage. At the stage of crystallization of silicates, which are the main carriers of strength, the rate of velocity growth reduces.

The initial stage of coagulation structure forming is characterized by horizontal sections, which end by inflection point (white point). It corresponds to the initial setting time (τ_1). Increasing in MK content leads to the intensifying of the ultrasonic pulse velocity.

This research method is more informative for the initial structure forming rather than setting time test. The setting time test enables us to estimate the coagulation period, whereas the ultrasonic pulse velocity test—to estimate the physical and chemical changes that lead to the paste's microstructure formation. However, the values of initial setting time determined by these both methods are correlated with each other. Introduction of SP and MK retards the initial structure formation due to SP adsorption films, which prevent hydration.

As can be seen, the curves of the ultrasonic waves' velocity passing through cement pastes from time to time contain areas that also characterize the structure formation stages: at the initial stage (induction period) C_1 increases slightly (horizontal section is noted); further, due to formation of low-strength, but significantly affecting C_1 crystallization framework of the aluminate components of cement $dC_1/d\tau$ gets maximum value. As the silicates crystallization starts, which are the main carriers of strength, the growth rate of C_1 decreases. The initial period of the coagulation structure formation on the ultrasonic velocity curves is characterized by a horizontal section ending with an inflection point (white dots) corresponding to time τ_1. The increase in the metakaolin content leads to the intensification of the passage of ultrasonic waves.

4.3.3 Peculiarities of Hydration and Structure of Hardened Cement Paste

Theoretical analysis permits us to suppose that the introduction metakaolin and superplasticizer into cement paste should affect the kinetics of cement hydration, phase composition and features of the hardened cement paste pore structure. Crystallization structure stage is characterized by the intensive strengthening of paste with SP and MK due to the active binding of portlandite with metakaolin and the formation of additional quantities of stable low-basic crystalline hydrate CSH (I). For more detailed testing of structure formation processes, X-ray and SEM analysis were applied. The influence of MK and SP on porous structure of cement paste and its phase composition had been determined.

Hydration Degree

The hydration degree kinetics actually characterizes the amount of chemically bound water in the cement hydration process and effect of admixtures on it. This characteristic was determined according to the method described in [115], at a different age of hardening according to the following formula:

$$\alpha = K_b^\tau \cdot 100/K_b \qquad (4.23)$$

where K_b^τ is the weight of bound water in a certain age of hardening, g;

K_b is the weight of bound water under complete hydration of cement, g.

The results of cement hydration degree are presented in Fig. 4.25.

Complete cement hydration is possible at its hardening in isolated space at W/C $\geq$ 0.5, and at water hardening $\geq$ 0.42 [116]. When W/C (W/B) is less than 0.42, the maximum hydration degree correlates to the initial W/C ratio [116] as follows:

$$\alpha = 2.38 \cdot 100 \cdot \text{W/C}. \qquad (4.24)$$

Substituting the corresponding value of W/C into Eq. (4.24), leads to $\alpha = 78.5\%$. However, the formula does not consider the impact of admixtures on hydration degree.

As can be seen from Fig. 4.23, increase in metakaolin dosage has a positive effect on the cement paste hydration degree, which is especially remarkable at the early stages of cement paste hardening. This phenomenon is explained by active binding of excess free lime, as well as by the performance of metakaolin particles as a substrate for reaction products crystallization [6, 117]. Along with the direct chemical interaction of metakaolin with cement, there is more intensive hydration of clinker minerals by themselves [6]. However, the partial substitution of cement by a high dosage of metakaolin (about 15%) may adversely affect the cement paste hydration degree due to the possible surrounding of new phases surface and prevention of contacts formation between the crystalline hydrates.

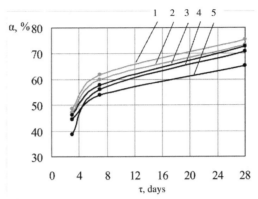

Figure 4.23. Impact of SP and MK content on Kinetics of Cement Hydration: 1) MK=10%, SP= 0.64%; 2) MK=15%, SP= 0.95%; 3) MK=5%, SP=0.46%;4) MK =0%, SP=0.32%; 5) MK=0%, SP=0%.

The adsorption film formed by the superplasticizer is sufficiently permeable to molecules of water and does not interfere with cement hydration and formation of a strong cement paste structure. Superplasticizer molecules cause disaggregation of cement flocs, improving water access to cement grains.

Porosity

The influence of MK and SP on pore structure of cement paste was studied. One of the most simple and universal methods of investigation of the pore structure of cement is the water absorption method [116]. It allows us to determine the integrated (apparent) porosity and differential parameters of pore structure (the average pore size and uniformity of pore size). Numerous experiments of water absorption of multicapillary materials such as hardened cement paste and concrete, can be described by exponential function. In general, water absorption curves of hardened cement pastes are approximated by three-parameter exponential function of the following form:

$$W_\tau = W_{max} \cdot (1 - e^{-\lambda \tau^\alpha}) \qquad (4.25)$$

where W_τ is water absorption after τ hours, %;

W_{max} is conventional value of maximum water absorption, %;

λ is a coefficient, which determines average diameter of capillary pores;

α is a coefficient, which determines uniformity of capillaries size;

τ is time of saturation of specimen, h.

To determine the porosity parameters according to the kinetics of water absorption, cubic specimens with 7.07 cm edge have been moulded. After curing at normal hardening conditions for 28 days, the specimens were dried at $t = 105...110°C$. The water absorption was determined several times starting from 2 h and till 24 h of saturation (which corresponds to conventional total open porosity). Kinetics of water absorption is shown on Fig. 4.24.

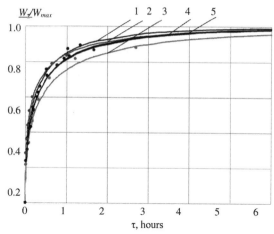

Figure 4.24. Kinetics of Water Absorption of Cement Pastes: 1 – MK=0%, SP=0%; 2 – MK =0%, SP=0.32% ;3– MK =5%, SP=0.46%; 4 – MK=10%, SP=0.64%; 5 – MK =15%, SP=0.95%.

For all the pastes (see Table 4.17) Eq. 4.25 is adequate at 95% confidence probability.

Composition 1: F_d=1.94<F_t=4.19; composition 2:F_d=1.77<F_t=4.19;

composition 3: F_d=2.05<F_t=4.53; composition 4:F_d=1.66<F_t=3.84;

composition 5: F_d=1.83<F_t=4.1.

The following parameters were calculated:

- weight water absorption, %:

$$W_m = \frac{m_{24} - m_0}{m_0} \cdot 100 \qquad (4.26)$$

- volumetric water absorption (open porosity), %:

$$W_v = \frac{m_{24} - m_0}{m_{24} - m_{24}^0} \cdot 100 \qquad (4.27)$$

where m_0 is the specimen weight in dry state (weight in air), kg;

m_{24} is weight the specimen after 24h of saturation (weight in air), kg;

is the specimen weight after 24h of saturation, (weight in water), kg.

Knowing the hardened cement paste hydration degree, its relative density, total porosity, capillary and gel porosity are obtained according to the calculated dependences [103]:

- relative density:

$$d = \frac{1 + 0.23 \cdot \alpha \cdot \rho_c /100}{1 + \rho_c \cdot W/C} \qquad (4.28)$$

- total porosity, %:

$$P_t = \rho_c \frac{\left(W/C - 0.23\alpha/100\right)}{1 + \rho_c \cdot W/C} \tag{4.29}$$

- gel porosity, %:

$$P_g = \frac{0.19 \cdot \alpha \rho_c / 100}{1 + \rho_c \cdot W/C} \tag{4.30}$$

- capillary porosity, %:

$$P_c = P_t - P_g = \frac{\rho_c \left(\dfrac{W}{C} - 0.42\dfrac{\alpha}{100}\right)}{1 + \rho_c \cdot \dfrac{W}{C}} \tag{4.31}$$

where α is hydration degree of cement (binder), %;

ρ_c is density of cement, $\rho_c = 3.1$ g/cm^3.

Taking into consideration that cement hydration significantly retards in the late age, the maximum hydration degree as that at the 28 days was used. Due to adding MK and SP the W/B ratio is considered constant. Experimental and calculated values of the cement paste pore structure parameters are given in Table 4.18.

Table 4.18. Parameters of Hardened Cement Paste Porous Structure.

No.	Admixtures, %		α	λ	W_m, %	W_v, %	D	Porosity		
	SP	MK						P_t, %	P_g, %	P_k, %
1	0	0	0.494	1.889	9.80	19.39	0.724	27.59	18.98	8.61
2	0.32	0	0.417	1.847	9.61	18.72	0.744	25.58	20.64	4.94
3	0.46	5	0.491	1.690	7.44	14.55	0.752	24.77	21.31	3.46
4	0.64	10	0.476	1.663	7.52	14.01	0.760	24.03	21.92	2.11
5	0.95	15	0.451	1.401	7.06	13.20	0.751	24.95	21.17	3.78

As can be seen from Table 4.18, an increase in metakaolin content in cement paste under other stable conditions leads to decreasing the average pore size and increasing its homogeneity. The open porosity (water absorption by volume) tends to decrease, the relative cement paste density increases. This phenomenon suggests the assumption of the active role of metakaolin in the hydration processes with the formation of fine-grained hardened paste with the advantage of stronger and more stable low-base calcium silicates type CSH (I) [81, 118].

According to the general laws of cement paste and concrete structure formation [81], the introduction of ultrafine materials leads to a change in the balance between gel ($1 \cdot 10^{-3} \leq d \leq 5 \cdot 10^{-3}\mu$m) and capillary ($5 \cdot 10^{-3} < d \leq 2 \cdot 10\ \mu$m) pores in favor of the gel, and the hardened cement paste structure becomes more dispersed. This is confirmed by calculated porosity data (Table 4.18, Fig. 4.25).

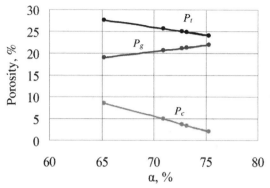

Figure 4.25. Dependence of total P_t, capillary P_c and gel porosity P_g of cement paste on hydration degree.

Compressive strength

There is also a correlation between the cement hydration degree and the cement paste compressive strength. A. Sheikin proposed the following correlation between these parameters [116]:

$$f_c = 310 \cdot \left(\frac{1 + 0.23 \cdot \alpha \cdot \rho_c / 100}{1 + W/C \cdot \rho_c} \right)^n \tag{4.32}$$

where $n = 2.7$.

Figure 4.26 shows the mean experimental values of the compressive strength for the studied cement pastes at 28 days and the corresponding hydration degree values. Cement paste specimens of 2 x 2 x 2 cm size hardened before testing in water. When approximating the data by the least squares' method, the experimental points can be described by Eq. (4.32) at $n = 4.1$ (Fig. 4.26). The equation is adequate with a 95% confidence probability: $F_d = 4.46 < F_t = 5.41$.

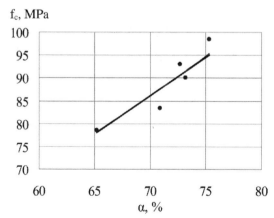

Figure 4.26. Dependence of Compressive Strength of Cement Paste on Hydration Degree.

X-ray Diffraction

To confirm the hypotheses regarding the reaction products chemical and mineralogical composition, X-ray study of cement paste at 28 days was performed. At the early structure formation stages of cement systems with ultrafine materials, a so-called "physical factor" dominates. It is expressed in filling the volume between the coarse cement particles and formation of numerous low-strength coagulation contacts [81]. The "chemical factor", which is expressed in the balance change between hydration products in the direction of a growing the number of stronger and more stable compounds, prevalent in the later periods of hardening.

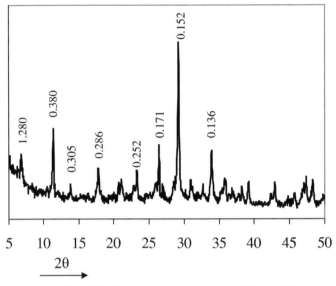

Figure 4.27. XRD of Composition "metakaolin-Ca(OH)₂" at 28 Days.

Due to the fact that metakaolin as a pozzolanic admixture interacts primarily with free lime — a product of clinker minerals hydration, samples of "Metakaolin—Ca(OH)$_2$" composition, 1: 1 by weight at a water-solid ratio of 1 (Fig. 4.27) were studied at 28 days. The results of X-ray analysis indicate the presence of the following minerals in the composition: calcite $CaCO_3$ (d = 0.305; 0.192; 0.171; 0.152; 0.136 nm), hydrogelenite (strathlingite) $2CaO·Al2O_3·SiO_2·8H_2O$ (C_2ASH_8, d = 1.280; 0.636 nm)—a mineral that has low strength by itself, however it is stable over time, and monocarbonate calcium hydroaluminate $3CaO·Al_2O_3·CaCO_3·11H_2O$ (d = 0.380; 0.286; 0.252 nm).

The results of X-ray diffraction (XRD) study of hardened cement paste at 7 and 28 days indicate the presence of partially crystallized tobermorite-like calcium hydrosilicate type CSH (I) of variable structure: (1–1.5) $CaO·SiO_2·H_2O$ (d = 0.305; 0.279; 0.210; 0.184 nm) in all the compositions (Fig. 4.28a,b). It should be noted that with increasing of the metakaolin portion in the binder and

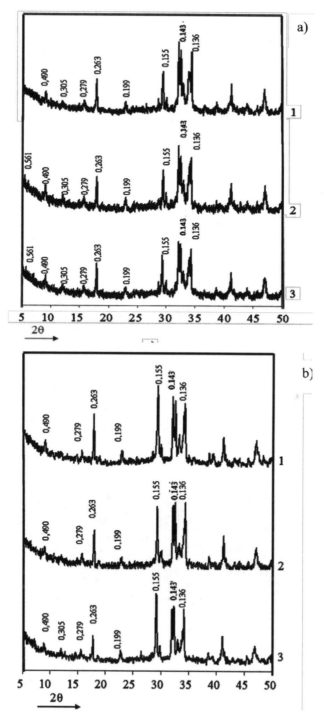

Figure 4.28. XRD of Cement Paste at 7 days (a), at 28 days (b): 1 – SP=0.32%; 2 – SP=0.45%, MK=5%; 3 – SP=0.64%, MK=10%.

hardening time, the minerals content increases, as it follows from the growth of the diffraction peaks intensity.

Systems also contain portlandite $Ca(OH)_2$ ($d = 0.490$; 0.263; 0.143 nm). Its amount decreases with increasing the metakaolin dosage. It should be also noted that the presence of a small amount of ettringite ($d = 0.561$; 0.155 nm) is available in the specimens both at early and late age of hardening. Its content is proportional to the amount of metakaolin and tobermorite-like calcium hydrosilicate CSH (II) $2CaO \cdot SiO_2 \cdot H_2O$ (0.280; 0.199 nm).

SEM Analysis

Results of SEM cement paste analysis (Fig. 4.29) indicate the predominance of fine-grained crystalline hydrates and gel-like reaction products. In general, the hardened cement paste structure becomes more dispersed. Thus, the results of X-ray diffraction and SEM analysis confirm the hypotheses about the active role

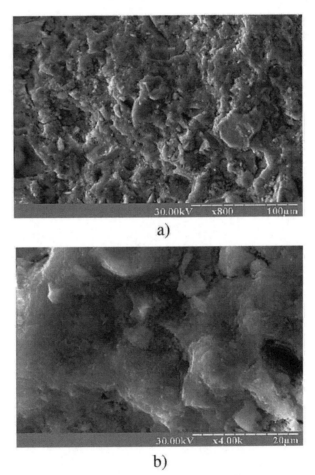

a)

b)

Figure 4.29. Microstructure of Hardened Cement Paste: SP=0.64%; MK=10% (a) ×800; b) ×4000.

of metakaolin in changing the balance between hydrated products in the direction of increasing the content of stronger and more stable fine crystalline hydrates type CSH(I), which causes redistribution of hardened cement paste porosity and strength. An increase in the amount of ettringite is expected to cause a decrease in the hardened cement paste shrinkage deformation.

Drying Shrinkage

Due to raised binder content in self-leveling high-strength concrete as well as the high water demand of mineral admixtures, the issue of hardened cement paste and concrete shrinkage is important. Early age shrinkage cracking is one of basic reasons for the destruction of concrete and reinforced concrete structures [119, 120]. According to available results, the high-performance concrete impact of water-cement ratio within the range of 0.35–0.5 is insignificant, whereas the cement paste volume has a significant impact on shrinkage [91]. Other factors, influencing shrinkage, are cement mineralogical composition, gypsum content, terms of hardening, type and dosage of admixtures, type of aggregates, size of specimens and structures [90, 121].

Plastic shrinkage develops in freshly mixed cement paste and concrete during the first few hours, at moulding until the end of setting and can result in the cracking of concrete and reinforced concrete structures. Developing of the shrinkage deformations during hardening are higher compared to plastic shrinkage. The basic reason for these deformations is drying and autogenous shrinkage to a lesser extent [122].

At hardened cement paste drying capillary-bound water evaporates initially, after that moisture adsorbed on the crystals surface evaporates [116]. Later, water between the layers of calcium hydrosilicate crystals from the tobermorite gel is disposed. Partial loss of chemically bound water is also possible. The initial shrinkage of cement paste at high relative humidity is caused by capillary forces. The lower the relative humidity is, the smaller the radius of the capillaries is.

Studies on the influence of metakaolin on autogenous shrinkage of hardened cement paste at a constant water-binder ratio shows that the shrinkage deformations increase at MK content up to 10% by cement weight. Shrinkage deformations decrease for specimens containing more than 15% of metakaolin [121]. Studies of the drying shrinkage of cement pastes with metakaolin content up to 15%, at 3 and 28 days of water hardening with subsequent air hardening at 20% and 55% showed that MK slightly increases drying shrinkage at an early age up to 3 days and reduces shrinkage deformations at 28 days. Reduction of shrinkage deformations at adding MK can be explained by the overall decreasing of water evaporation in the hardened cement paste, as it has a low-porosity structure with pores of smaller diameter due to the filling effect of MK and its pozzolanic reactions [123].

The main products of MK pozzolanic reactions are low-base calcium hydrosilicates of (C-S-H) and hydrogehlenite (C_2ASH_8). In this case, an increase in the metakaolin portion in binder leads to a decrease in the paste autogenous shrinkage at W/B = 0.3 [124]. The reason for autogenous shrinkage of high-performance concrete with W/B <0.4 is considered as the lack of excessive water for evaporation and formation of capillary pores [125]. An important factor influencing its determination is the coating of the specimens' surface with an impermeable layer [102]. At that point the contraction of reaction products depends on the temperature and relative humidity after finishing plastic deformation during the first weeks of hardening. In the case of self-leveling and self-compacting concrete at equal W/C and water content, addition of mineral admixtures in superplasticized concrete leads to a reduction in shrinkage deformation at drying under 20°C and a relative humidity of 65–70%. Most studies prove the positive effect of MK on the reduction of drying shrinkage as well as contractile shrinkage [1, 126, 127].

Equations for predicting drying shrinkage often take into account the impact of W/C [90]. It is also believed that calculation according to the CEB-FIB model gives underestimated values of shrinkage compared to the experimental ones, and according to the ACI model—overestimated [128]. Existing formulas mainly offer the estimation of ultimate shrinkage deformations [129]. It is also possible to calculate the shrinkage values of hardened cement paste and concrete at a certain age [91]. Such equations do not take into account the characteristics of raw materials and the content of chemical and mineral admixtures. At the same time, the mathematical description of shrinkage deformations kinetics is significantly complicated [128].

For the reasons of homogeneity and comparability of the initial parameters and in order to eliminate the influence of the water-binder ratio as the most important factor, all samples were made on the basis of mixtures with water-binder ratio of 0.3. The flow spread diameter D_f was obtained for each batch using a cone of the Vicat apparatus.

To determine the shrinkage deformations, 4 × 4 × 16 cm prism specimens were moulded. The values of linear shrinkage were determined using a clock-type indicator. The initial count was taken as the value at 1 day of hardening immediately after demoulding the specimens. During the study period, all samples were stored in air-dry conditions. The test result was taken as the average value of deformations for the two specimens. The tests were performed after 90 days of hardening. Relative linear deformation of the specimens was measured (ε^{90},%).

Three series of experiments were conducted. At the first stage, the effect of the superplasticizer type and its dosage on the shrinkage deformations of hardened cement paste was studied (Table 4.19), at the second stage, the effect of metakaolin dosage on shrinkage was determined (Table 4.20). At the third stage, comparison of the metakaolin type (specific surface area) effect at the same dosage was performed (Table 4.21).

The type of superplasticizer and dosage affects the cement paste shrinkage deformation (Fig. 4.30). According to experimental data, significant dosages of SP, especially above the recommended values (SP_1), lead to retarding hardening process and intensive evaporation of moisture. However, variations of shrinkage

Table 4.19. Compositions and Test Results of Drying Shrinkage (1st set of experiments).

Specimen	1–1	1–2	1–3	1–4	1–5	1–6	1–7	1–8	1–9	1–10
SP type	SP_1	SP_1	SP_1	SP_1	SP_2	SP_2	SP_2	SP_3	SP_3	SP_3
%SP	0	0.25	0.5	0.75	0.125	0.25	0.375	0.1	0.2	0.3
D_f, cm	8	8.5	16	21.5	8.3	16.2	21.2	14.3	21.3	23.5
ε^{90}, %	0.231	0.244	0.287	0.287	0.203	0.190	0.197	0.206	0.102	0.189

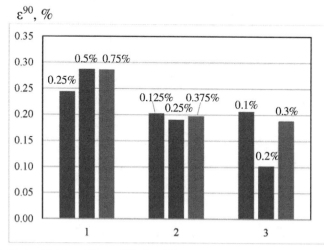

Figure 4.30. Drying Shrinkage Values of Hardened Cement Paste Depending on Superplasticizer Type and Dosage: $1 - SP_1$; $2 - SP_2$; $3 - SP_3$.

Table 4.20. Compositions and Test Results (2nd set of experiments, SP_1, MK_1).

Specimen	2–1	2–2	2–3	2–4	2–5		
%SP_1	0.5	0.75	1	1.25	1.5		
%MK_1	5	5	5	5	5		
D_f, cm	9.7	15	18.8	21,7	22.5		
ε^{90}, %	0.208	0.257	0.271	0.309	0.329		
Specimen		2–6	2–7	2–8	2–9	2–10
%SP_1		0.75	1	1.25	1.5	1.75
%MK_1		10	10	10	10	10
D_f, cm		10.7	14.7	17.8	19.7	21.3
ε^{90}, %		0.2338	0.2381	0.2597	0.265	0.3033
Specimen			2–11	2–12	2–13	2–14
%SP_1			1	1.25	1.5	1.75
%MK_1			15	15	15	15
D_f, cm			9,3	12,4	14,5	15,8
ε^{90}, %			0.213	0.237	0.278	0.288

values for the same SP type are not significant. The increase in shrinkage at the maximum SP$_3$content can be explained by the cement paste heterogeneity and limited water bleeding.

By adding metakaolin, the tendency to shrinkage growth at subsequent increasing SP content is observed for different dosages of mineral admixture (Table 4.20, Fig. 4.31). The trend is linear for each dosage of MK. Instead, the dependence of shrinkage deformations on the MK dosage is polynomial. The minimum is achieved for dosages of 10%, due to the peculiarities of the early structure forming of metakaolin-based pastes structure. The experimental data are then approximated by an equation with an approximation reliability value not less than 0.9:

$$\varepsilon = A \cdot X_1 + B \qquad (4.33)$$

where A= $0.0018\ X_{22} - 0.0375 \cdot X_2 + 0.2596$;

B = $-0.0018 \cdot X_{22} + 0.0305 \cdot X_2 + 0.0489$,

X_1 is the superplasticizer content, % by binder weight,

X_2 is the metakaolin content, % by binder weight.

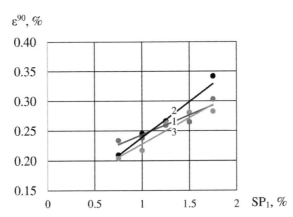

Figure 4.31. Drying Shrinkage Values of Hardened Cement Paste Depending on SP and Metakaolin
Dosage: 1 – MK$_1$=5%; 2- MK=10%, 3 – MK=15%.

Increasing the specific surface area of metakaolin (Table 4.21, Fig. 4.32) leads to a natural decrease in the paste's workability. At the same time, there is no clear correlation between the values of metakaolin specific surface area and shrinkage deformations of hardened cement paste samples.

Comparison of shrinkage deformations values for specimens of equal flowability pastes (flow spread diameter of Vicat apparatus D$_f$ = 20 ± 2 cm) is presented in Fig. 4.33. This data confirms the dominant role of the superplasticizer content on shrinkage.

The superplasticizer dosage has a more significant impact on the shrinkage deformations of the studied cement paste specimens at constant water-binder

Table 4.21. Compositions and Test Results (3rd set of experiments, SP_1, MK = 10%).

Specimens	3-1	3-2	3-3	3-4	3-5
% SP	0.75	1	1.25	1.5	1.75
Type of MK	MK_1	MK_1	MK_1	MK_1	MK_1
D_f, cm	10.7	14.7	17.8	19.7	21.3
ε^{90}, %	0.234	0,238	0,260	0,265	0,303
Specimens	3–6	3–7	3–8	3–9	3–10
%SP	0.75	1	1.25	1.5	1.75
Type of MK	MK_2	MK_2	MK_2	MK_2	MK_2
D_f, cm	8.8	13	15.5	17.2	18.5
ε^{90}, %	0.209	0.245	0.267	0.281	0.342
Specimens	3–11	3–12	3–13	3–14	3–15
%SP	0.75	1	1.25	1.5	1.75
Type of MK	MK_3	MK_3	MK_3	MK_3	MK_3
D_f, cm	8	9.2	10	12	13.5
ε^{90}, %	0.204	0.217	0.265	0.281	0.283

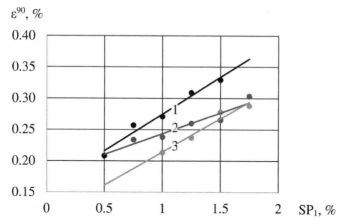

Figure 4.32. Drying Shrinkage Values of Hardened Cement Pastes Depending on SP content and MK Type: 1 – MK_1; 2 -MK_2; 3 – MK_3.

ratio. At adding metakaolin, this dependence is linear. The change in the specific surface area of metakaolin does not have a significant effect on shrinkage. The optimal metakaolin content in terms of minimizing shrinkage is 10% by binder weight. It can be explained by the high water retaining capacity of metakaolin, which prevents the rapid surface drying of the specimens and reduces the total volume of cement.

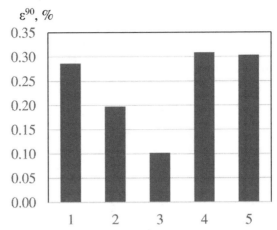

Figure 4.33. Drying Shrinkage Values of Hardened Cement Pastes at Equal Flow Spread Diameter:
1 – SP_1=0.75%; 2 – SP_2=0.375%; 3 – SP_3=0.2%; 4 – SP_1=1.25%; MK=5%; 5 – SP_1=1.75%; MK=10%.

4.4 Properties of Self-leveling Concrete Containing Metakaolin

Research of cement-based systems containing metakaolin continued with studying the properties of self-leveling concrete close by its workabilty to self-compacting one [130]. The impact of metakaolin and other technological factors on workability, superpalasticizer dosage, water bleeding and water retention, air content of fresh concrete had been studied.

4.4.1 Water Consumption and Workability of Fresh Concrete

Approaches of Workability Retention Measuring

Self-compacting and self-leveling high-strength concrete has a high cement content and relatively low values of W/C (W/B). Variations in cement content do not have a significant impact on concrete workabilty within the scope of the water consumption constancy rule, i.e., at cement content up to 350 kg/m^3 [88] or, according to other data, up to 400 kg/m^3 [131] and at least 40% of mortar volume [132] .The growth of cement content typically does not make any influence on the workability of fresh concrete. However, at a higher cement consumption, which is typical for high-strength concrete (over 400 kg/m^3), the rule is not applied and the workability decreases. The use of multifunctional modifiers based on superplasticizer and metakaolin compensates for the water demand increase caused by adding the mineral admixture in concrete and even reduce it.

The superplasticizer efficiency increases with the cement paste volume in concrete [81]. At dosages of admixtures that exceed the so-called" efficiency threshold" (it is in the range of 0.9 ... 1.5% by cement weight for superplasticizer (SP) of naphthene formaldehyde (NF) type), the cement paste viscosity does not change, and the change in the mixture water consumption is already insignificant [86]. However, when using a superplasticizer in combination with a finely dispersed mineral admixture to achieve the same workability of fresh concrete, the application an excessive amount of superplasticizer is required, which in some cases may exceed the efficiency threshold (see subsection 4.3) [86].

An important issue in the study of the mineral admixtures self-leveling effectiveness is the prediction of workability decreasing with time, the so-called "liveability" [86], or workability retention of fresh concrete over time [133, 134]. It allows us to make adjustments in fixing the initial workability and determine the allowable transportation duration of fresh concrete. The fresh concrete workability retention depends on the temperature and humidity of the environment, the mineralogical and chemical composition of cement, the mineral admixtures type and dosage and chemical admixtures used [101, 111]. Currently, there is no clear criteria for assessing the workability retention. Thus, to predict the possible value, the concepts of "setting time of fresh concrete" and formulas for determining the initial setting time of fresh concrete had been proposed [134–137].

According to Ukrainian standards, the workability retention is measured each 30 min during the time range determined by the manufacturer, and different samples of fresh concrete are used for each test. The procedure is time consuming and impractical in case of self-leveling concrete, as its workability before placing should be provided without additional compaction or minor compaction. Testing the workability during 1.5–2 h after mixing in some cases does not allow to fix slump change when it occurs within the same slump range of one class [86].

There was a suggestion to apply time of change either from the upper to the lower limit of the given slump class or from the average value of a given class to the average value of the lower neighboring class [133] as the retention criterion. We have considered the second option as a more convenient one for self-leveling concrete. It was also shown that the determination of the workability retention can be performed at a single batch of the fresh concrete, as frequent mixing, which improves workability, compensates an increase in the binder particles activation and hydration processes intensification. This method does not require special equipment except traditional Abrams cone. Thus, for the fresh concrete of slump class S5 the workability indicator represents time of decreasing the average slump from 22 to 18 cm:

$$\tau^r_{22-18} = \tau_{22} - \tau_{18} \qquad (4.34)$$

All batches studied below had an initial slump of (22 ± 1.5) cm, the interval of testing was 30 min. SP_2 (NF type with defoaming agent) was used. Measurements were stopped if the last mean value of slump was less than 18 cm. The mixtures composition was calculated according to a known method [88].

Results of Workability Retention and Discussion

The results of the first research stage give a comparative estimation of workability retention for self-leveling concrete (Table 4.22).

Table 4.22. Concrete Compositions for Determination the Workability Retention

No.	W/C	W/B	Basic components, kg/m³				Admixtures, % of binder weight / kg/m³	
			Water	Cement	Sand*	Crushed stone	SP	MK
1	0.42	0.42	260	618	451	1000	–	–
2	0.40	0.40	200	500	610	1105	0.50 / 2.5	–
3	0.44	0.40	200	450	606	1097	0.75 / 3.75	10 / 50

Note. Mixture of fine aggregates S_1 and S_2 was applied (ratio 1:1, fineness modulus = 2.0)

Results of the workability retention kinetics of fresh concrete are shown in Fig. 4.34. As can be seen from the figure, the fresh concrete slump dependence on time can be described by a quadratic equation. The corresponding equations, adequate at 95% confidence probability are:

for composition 1: $Sl = -1.7\tau^2 - 1.0\tau + 22.0$, $F_d = 2.70 < F_t = 19.30$;

for composition 2: $Sl = -2.3\tau^2 - 2.2\tau + 22.0$, $F_d = 2.59 < F_t = 19.30$;

for composition 3: $Sl = -0.9\tau^2 + 0.6\tau + 22.0$, $F_d = 2.17 < F_t = 8.94$.

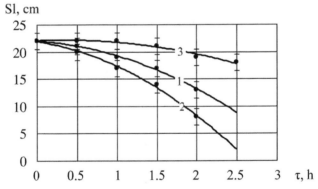

Figure 4.34. Workabilty Retention of Fresh Concrete: 1 – control; 2 – SP=0.25%; 3 – SP=0.75%, MK=10%.

According to the experiment conditions $\tau_{22} = 0$ for all batches; then τ_{18} was determined by the substitution of the corresponding slump value (18 cm) to the equation of the approximate curve.

The fresh concrete with SP has highest workability loss over time, concrete with addition of superplasticizer and metakaolin has the lowest one. There are experimental data that confirms that concrete with superplasticizer loses workability faster than that of the same control batch, which is naturally explained by the higher water content of the latter [111]. Adding superplasticizer and metakaolin retards the processes of the initial structure formation and the slump loss within 2 h is insignificant. Thus, the workability retention of concrete samples 1–3 is 1.27 h (or 75 min), 0.92 h (55 min) and 2.45 h (147 min), respectively.

It should be noted that kinetics of workability retention correlates with that of cement pastes conductivity at the initial stage of structure formation (see subsection 4.3.2), which confirms the accuracy of the hypotheses.

Planned Experiment and Workability Retention Research

At the next research stage, in order to determine the effect of metakaolin on workability retention over time and superplasticizer dosages required to achieve their required workability of S5, an experiment was carried out using a four-factor three-level plan close to D–optimal [138, 139]. Experiment planning conditions are given in Table 4.23.

Planned experiment matrix, concrete compositions for each batch, experimental values of SP content and workability retention are shown in Table 4.24.

Regression equations of workability retention adequate at 95% confidence probability have been obtained:

$$\tau'_{22-18} = 2.35 + 0.22x_1 - 0.18x_2 + 0.19x_3$$
$$+ 0.14x_4 - 0.11x_1^2 - 0.07x_2^2 + 0.06x_4^2 - 0.05x_2x_3 \qquad (4.35)$$
$$F_d = 2.40 < F_t = 8.81$$

As can be seen from Fig. 4.35, under the condition of constant W/B, increase in the metakaolin portion in binder from 5 to 15% leads to slump retention growth of about 20 minutes. As can be seen from Eq. (4.37), factors x_1 and x_3 have the highest effect on the retention parameter. It is known that W/C (W/B) significantly affects the change in workability: according to [140], at increased W/C, a shielding film of C_3SH_x—the product of C_3S hydration—forms, which retards further cement hydration. During mixing, this film dissolves and the hydration process intensifies. The growth of metakaolin dosage in binder leads to an increase in workability retention at any binder consumption within the variation range (Fig. 4.35). It can be explained by the fact that metakaolin particles, adsorbing water and SP, initially inhibit hydration processes. After

Table 4.23. Conditions of Experiment Planning.

Factors designation		Variation level		
Designation	Name of factors	Bottom (–1)	Mean (0)	Upper (+1)
X_1	Water-binder ratio (W/B)	0.34	0.37	0.40
X_2	Binder content (B), kg/m³	450	500	550
X_3	Metakaolin portion (MK), % by binder content	5	10	15
X_4	Fineness modulus of sand (M_f)	1.6	2.0	2.4

Table 4.24. Planned Experiment Matrix, Concrete Compositions and Experimental Values.

No.	Natural values of factors				Components content, kg per m³ of concrete						τ_{22-18}, h
	x_1 (W/B)	x_2 (B)	x_3 (MK)	x_4 (M_f)	W	PC	S	CS	MK	SP, %	
1	0.40	550	15	2.4	220	467.5	543	1070	82.5	0.55	2.75
2	0.40	550	15	1.6	220	467.5	537	1078	82.5	0,75	2.42
3	0.40	550	5	2.4	220	522.5	547	1078	27.5	0.20	2.08
4	0.40	550	5	1.6	220	522.5	541	1086	27.5	0.50	1.83
5	0.40	450	15	2.4	180	382.5	663	1147	67.5	1.20	2.92
6	0.40	450	15	1.6	180	382.5	657	1156	67.5	1.45	2.58
7	0.40	450	5	2.4	180	427.5	666	1153	22.5	0.51	2.67
8	0.40	450	5	1.6	180	427.5	660	1162	22.5	0.75	2.33
9	0.34	550	15	2.4	187	467.5	573	1129	82.5	1.32	2.08
10	0.34	550	15	1.6	187	467.5	567	1138	82.5	1.65	1.92
11	0.34	550	5	2.4	187	522.5	577	1137	27.5	0.55	1.75
12	0.34	550	5	1.6	187	522.5	571	1146	27.5	0.80	1.58
13	0.34	450	15	2.4	153	382.5	690	1193	67.5	2.10	2.50
14	0.34	450	15	1.6	153	382.5	683	1202	67.5	2.26	2.25
15	0.34	450	5	2.4	153	427.5	693	1199	22.5	1.00	2.17
16	0.34	450	5	1.6	153	427.5	686	1209	22.5	1.30	1.92
17	0.40	500	10	2	200	450	606	1113	50	0.75	2.50
18	0.34	500	10	2	170	450	634	1166	50	1.35	2.00
19	0.37	550	10	2	204	495	562	1103	55	0.70	2.17
20	0.37	450	10	2	167	405	662	1192	45	1.15	2.42
21	0.37	500	15	2	185	425	618	1136	75	1.48	2.50
22	0.37	500	5	2	185	475	622	1143	25	0.75	2.17
23	0.37	500	10	2.4	185	450	626	1131	50	0.95	2.58
24	0.37	500	10	1.6	185	450	611	1148	50	1.20	2.25

Note. W – water, PC -Portland cement, S – sand; CS – crushed stone, SP – superplasticizer, MK – metakaolin.

re-mixing, the hydration processes accelerate. It is noteworthy that there is an interaction between factors x_2 and x_3: thus, the fresh concrete workability retention at the binder content changes varies more significantly at a high consumption of metakaolin (15% by weight of binder) rather than at low one. At the same time, increase in the binder content leads to an intensive adsorption of superplasticizer on cement and metakaolin grains. In its turn, it leads to workability decrease. At reduction of sand fineness modulus, i.e., with an increase in the surface of particles to be mostified, there is a decrease in the retention parameter. Increasing in the metakaolin consumption does not change the curves nature at any fineness modulus value in the studied range of factors variation. Therefore, increasing

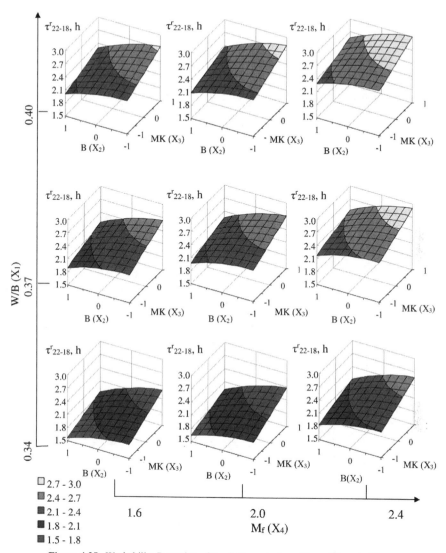

Figure 4.35. Workability Retention of Fresh Concrete from Slump 22 cm to 18 cm.

of metakaolin content leads to a retardation of initial structure formation and a corresponding retention of concrete workability.

Influence of the Factors on Superplasticizer Dosage

To maintain the constant water consumption (W/B) it is necessary to increase the dosage of superplasticizer at adding metakaolin. Based on the experimental data shown in Table 4.24, were the obtained regression equation of the SP dosage, % by weight of the binder depending on the factors given in Table 4.23:

$$SP_{NF} = 1.04 - 0.32x_1 - 0.26x_2 + 0.36x_3 - 0.13x_4 - 0.11x_2^2 + 0.08x_3^2 - 0.11x_1x_3 - 0.08x_2x_3$$

$$F_d = 2.09 < F_t = 8.81. \tag{4.36}$$

Figure 4.36 shows the diagrams of SP of NF type dosage depending on given factors. As can be seen from Fig. 4.36, at high metakaolin consumption (15%) the change of W/B has a more significant effect on the SP content than at a low consumption. Following Eq. (4.36), an increase in the metakaolin dosage and W/B most significantly affects the growth of superplasticizer content. Thus, when replacing 5% of cement by metakaolin, the change in W/B from 0.34 to 0.4 leads to an increase in the SP consumption from 0.54 to 1.05%, whereas replacing 15%, leads to an increase in the SP consumption from 1.0 to 1.9%

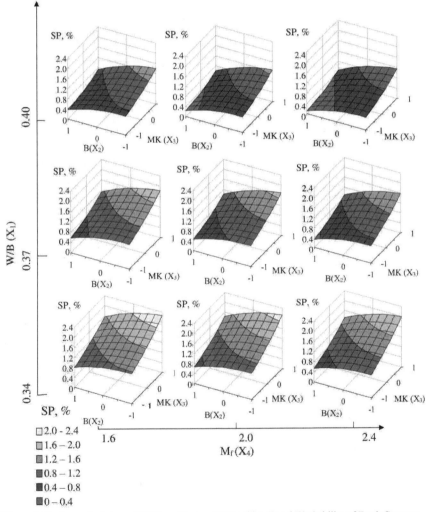

Figure 4.36. Superplasticizer of NF Type Dosage for Reaching Equal Workability of Fresh Concrete.

by the binder weight. Figure 4.36 indicates the higher sensitivity of concrete with higher binder content to the changes in SP dosage rather than concrete with a lower cement content and low W/B. Thus, when the metakaolin portion increases from 5 to 15%, the SP content increases from 0.45 to 1.0% at a binder consumption of 550 kg/m^3, and from 0.83 to 1.7% when binder consumption is 450 kg/m^3. The increase in binder consumption leads to a decrease in SP NF dosage. This is consistent with the known data [86], according to which the effect of superplasticizer is more noticeable in concrete with a higher cement consumption. The growth of sand fineness modulus leads to a natural increase in the superplasticizer content, as it increases the fresh concrete water demand.

Thus, the combined adding of superplasticizer and metakaolin enables us to obtain self-leveling concrete with a more stable workability over time rather than control concrete and concrete with SP admixture.

For comparing the effect of SP of NF type and PC type, research was conducted according to three-level three-factors plan, similar to a D-optimal one [139, 141]. Quartz sand with fineness modulus 1.5–1.6 was used for the research (Fig. 4.39).

According to Table 4.25, SP of polycarboxylate (PC) type ranges between 0.16 and 0.51%, so the dosages are comparatively lower than for SP of NF type. It demonstrates that PC SP yields an extra plasticizing effect on fresh concrete and significantly reduces the concrete water demand at lower doses, comparing

Table 4.25. Matrix of Planning and Concrete Compositions (SP of PC type).

Test No.	Matrix of planning			Components content, kg per m^3 of concrete					
	W/B X_1	B, kg/m^3, X_2	MK, %, X_3	PC	W	S	CS	MK	SP(NF)
1	0.4	550	15	468	220	537	1078	82	1.38
2	0.4	550	5	523	209	551	1106	27	0.89
3	0.4	450	15	383	153	683	1202	67	1.73
4	0.4	450	5	428	171	669	1177	22	1.17
5	0.34	550	15	468	159	592	1189	82	1.77
6	0.34	550	5	523	178	579	1162	27	1.49
7	0.34	450	15	383	130	706	1242	67	2.30
8	0.34	450	5	428	145	694	1222	22	1.76
9	0.4	500	10	450	180	616	1157	50	1.15
10	0.34	500	10	450	153	641	1205	50	1.80
11	0.37	550	10	495	183	572	1149	55	1.32
12	0.37	450	10	405	150	688	1211	45	1.53
13	0.37	500	15	425	157	636	1194	75	1.57
14	0.37	500	5	475	176	622	1168	25	0.97
15	0.37	500	10	450	167	629	1181	50	1.25
16	0.37	500	10	450	167	629	1181	50	1.29
17	0.37	500	10	450	167	629	1181	50	1.35

to SP of NF type. That is caused by the steric (spatial) effect of SP PC type on the cement particles repulsion and fresh concrete dilution prevailing under an electrostatic one that is typical for sodium naphthalene formaldehyde (NF) type [142]. The linear terms coefficients in Eq. (4.37) have similar absolute values. As well as for NF SP increasing in water-binder ratio (x_1) as well as in binder content (x_2) leads to SP content reduction. It meets the known data that the effect of superplasticizers is more notable at sufficient water content and higher cement content [143, 144].

The results of experiments are approximated by the following regression equation:

$$SP_{PC} = 0.26 - 0.06x_1 - 0.06x_2 + 0.05x_3 + 0.04x_1^2 + 0.03x_2^2 \qquad (4.37)$$

The equation is adequate, $F_d = 4.31 < F_t = 19.38$.

The quadratic terms' coefficients are inherent only for factors x_1, x_2, thus the dependence form is close to linear one that is most evident at higher water-binder ratio (Fig. 4.37).

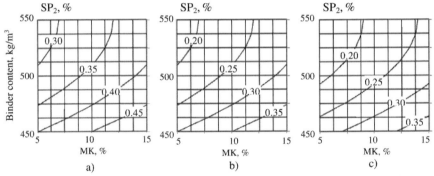

Figure 4.37. Influence of Binder Content (x_2) and Metakaolin Dosage in Binder (x_3) on SP PC Type Dosage: a) W/B = 0.34; b) W/B = 0.37; c) W/B = 0.40.

There was no significant interaction of factors when SP_3(superplasticizer of polycarboxylate type) (see subsection 4.2.2) was used. Following Fig. 4.37c, the binder content and metakaolin portion change have a lower influence on SP_2 consumption, compared to SP_1. The character of SP_{PC} dosage dependence on binder and metakaolin content is constant at different W/B values. As it can be followed from Fig. 4.37c, for W/B = 0.4 at binder content of 450 kg/m³ and metakaolin portion from 5% to 15% the SP dosage varies from 0.27 to 0.36% and for 550 kg/m³ binder content with the same MK the SP dosage varies from 0.16 to 0.25%.

4.4.2 Water Bleeding, Segregation and Air-entraining of Self-leveling Concrete

Important parameters of self-leveling concrete homogeneity are water bleeding and segregation. Concrete is a multicomponent system, and the cement sedimentation processes occuring from mixing till setting, are caused by various average diameters of concrete components. During these processes, water displaces or bleeds, the sedimentation degree depends on the size of cells in the frame of densified concrete structure [86].

The cement paste's ability to retain a certain volume of water depends on the interaction forces of its particles. The denser the cement particles are, the smaller the cells of the structural spatial grid are and the stronger the water retention of the cement paste is. As if self-leveling concrete has basal structure, the particles are spaced at considerable distances and are unable to interact with each other, the structural lattice is destroyed and the cement paste is bleeding. The separated water, being in a suspended state, gradually settles and a layer of water is formed over the settled cement [145].

For fresh concrete batches of equal workability, adding the superplasticizer reduces water separation, as water consumption becomes lower. However, danger of water separation exists even at introduction of superplasticizer [111]. According to available data [146, 147] when cement consumption is less than 400 kg/m^3 it is effective to use water-retaining mineral admixtures, such as bentonite clay, which has the ability for swelling in water (approximately 15 times) and to retain excessive water in the pores, like fly ash removal [111, 148], which increases the fresh concrete homogeneity without the significant reduction of workability.

Water bleeding is in direct relation to water retention of the cement paste and fresh concrete. As it was shown in subsection 4.3.1, the maximum value of water-binder ratio, which characterizes the maximum water retention of the paste in a static state with the composite admixture of NF type superplasticizer and metakaolin is 2.5 times higher than normal consistency, which is 80% higher than cement pastes without additives (1.65 times). Metakaolin, due to the developed particle shape, also binds water intensively, thus reducing the fresh concrete water bleeding. This statement is confirmed by the results of determining the water bleeding of fresh concrete of batches listed in Table 4.21.Water bleeding was obtained after settling the mixtures in glass graduated measuring cylinders and characterized by the amount of water that separated during 1.5 h of settling, % by volume.

Figure 4.40 presents a comparative diagram of self-leveling concrete water bleeding. As can be seen from the data presented, the introduction of metakaolin and superplasticizer (composition 3) reduces the concrete mixtures water separation compared to those without admixtures and with SP, due to a more developed surface and correspondingly higher water consumption of metakaolin-based fresh concrete.

Possibility of internal and external segregation was discussed [111, 146]. The first one is due to the action of grains gravitational force. It decreases with increasing the mortar viscosity or decreasing the aggregate grain size. The second one occurs as a result of insufficient crushed stone adhesion to the cement-sand mortar, which may be due to excessive mortar viscosity or increased content of crushed stone [105, 146]. The internal segregation is the characteristic of self-leveling concrete, whereas the external one can occur in fresh concrete of any workability[105].

The mortar segregation index, determined by standard requirements, characterizes the fresh concrete cohesion under dynamic action (after vibrocompaction for 25 sec). Self-leveling concrete as well as SCC, is not subjected to vibrocompaction and fills the volume provided to them by its own weight. However, during transportation and application, it is exposed to short-term vibration effects that can disrupt the concrete structure homogeneity. It should be mentioned, that flow spread diameter of the Abrams cone is a more relevant parameter of workability for self-leveling concrete and SCC with a slump more than 20 cm [90, 132].

A method for rapid assessment of fresh concrete cohesion and segregation using the tangent of the external friction angle is suggested. It is determined by the slump and the flow spread diameter of the Abrams cone (Fig. 4.38):

$$tg\varphi = (30 - Sl)/(0.5D_f) \tag{4.38}$$

where Sl and D_f—are the slump and the flow spread diameter of the cone, respectively, cm. Studies have shown that the values of this indicator correlate with the corresponding standard of segregation parameter. To achieve a mortar segregation less than 5%, the value of $tg\varphi$ should be ≥ 0.25 [87].

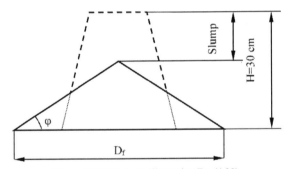

Figure 4.38. Diagram Illustrating Eq. (4.38).

Figure 4.39 shows the results of determining the mortar segregation indicators by standard method (b) and Eq. (4.38) (c). As can be seen here, there is some discrepancy between the data for concrete samples 1 and 2. This can be explained by the fact that the mixtures without mineral admixture have both internal and external segregation. For batch 1 it is caused by a high volume of cement paste, which increases the concrete mixture cohesion and prevents its spreading. In the case of adding superplasticizer, the mixture has a significant amount of coarse aggregate, which can cause external segregation. The indices of tg φ and segregation given for composition 3 practically correspond to the values given in [87]. Introduction of metakaolin allows us to obtain mixtures without evident signs of external segregation. At the rational concrete proportioning (limitation

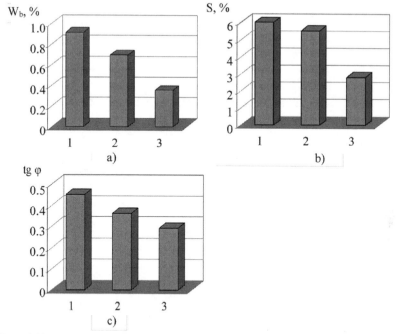

Figure 4.39. Parameters of Water Bleeding W_b (a), Segregation S (b) and tgφ (c) of Self-leveling Concrete: 1– control; 2– SP=0.25%; 3– SP=0.75%, MK=10%.

of crushed stone consumption, absence of cement overconsumption), there can be observed only an internal segregation due to increasing viscosity of the mortar component at adding metakaolin.

Workability parameters, segregation index tgφ and the entrained air volume of fresh concrete were determined. The experiment was implemented according to a four-factor three-level plan (see Table 4.22). The research results are given in Table 4.26.

For all the batches water bleeding has not exceed 0.2%. It indicates high homogeneity of fresh concrete. Therefore, obtaining special quantitative dependences for regulation of this parameter was not required in the research. The batches have no signs of segregation, estimated by tg φ parameter.

As a result of the experiment, the following regression equation for tg φ was obtained:

$$\text{tg } \varphi = 0.32 - 0.06\,x_1 - 0.03x_2 + 0.04x_3 - 0.04x_4 + 0.02x_1^2$$
$$+ 0.03x_3^2 - 0.02x_1x_2 + 0.02x_1x_4 \tag{4.39}$$
$$F_d = 6.73 < F_t = 8.89.$$

Figure 4.40 shows dependencies that reflect the influence of the studied factors on tg φ. As can be seen from Eq. (4.41) and Fig. 4.40, an increase in the water-binder ratio reduces the fresh concrete cohesion and leads to decreasing tg

Table 4.26. Planned Experiment Matrix and Experimental Values.

No.	Matrix of planning				Experimental values			
	X_1 (W/B)	X_2 (B)	X_3 (MK)	$X_4(M_f)$	Slump, cm	D_f, cm	$tg\varphi$	V_a, %
1	0.40	550	15	2.4	23	52	0.27	0.46
2	0.40	550	15	1.6	22	47	0.34	0.82
3	0,40	550	5	2.4	23	61	0.23	0.75
4	0.40	550	5	1.6	23	54	0.26	1.56
5	0.40	450	15	2.4	24	39	0.31	0.77
6	0.40	450	15	1.6	22	41	0.39	1.42
7	0.40	450	5	2.4	23	54	0.26	1.20
8	0.40	450	5	1.6	22	50	0.32	2.10
9	0.34	550	15	2.4	23	40	0.35	0.85
10	0.34	550	15	1.6	22	36	0.45	1.56
11	0.34	550	5	2.4	24	40	0.30	1.50
12	0.34	550	5	1.6	22	41	0.39	2.31
13	0.34	450	15	2.4	23	38	0.37	1.52
14	0.34	450	15	1.6	22	36	0.45	2.28
15	0.34	450	5	2.4	23	42	0.33	2.14
16	0.34	450	5	1.6	22	39	0.41	3.17
17	0.40	500	10	2	23	48	0.29	0.95
18	0.34	500	10	2	23	38	0.37	1.85
19	0.37	550	10	2	24	35	0.34	1.40
20	0.37	450	10	2	22	55	0.29	1.95
21	0.37	500	15	2	22	44	0.36	1.65
22	0.37	500	5	2	22	53	0.30	2.10
23	0.37	500	10	2.4	24	44	0.27	0.81
24	0.37	500	10	1.6	23	40	0.35	1.84

φ. Binder content growth leads to a slight increase in angle φ. It can be explained by the fact that the content of the binder-sand mortar remained almost the same. Increase in the metakaolin content leads to increase in the mortar cohesion and, accordingly, to the growth of tg φ. Thus, at W/B = 0.34, increasing metakaolin portion in the binder from 5 to 15% by weight causes growth of tg φ from 0.36 to 0.48 due to the increased fresh concrete viscosity. The binder content growth leads to increase in the flow spread diameter of the fresh concrete and, accordingly, to decrease in tg φ. Increasing in the sand fineness modulus leads to decrease in the value of φ. In general, values of tg φ in the research vary from 0.23 (W/B = 0.4; B = 550 kg/m³, MK = 5%, M_f = 2.4) to 0.45 (W/B = 0.34, B = 450 kg/m³, MK = 15%, M_f = 1.6). It confirms high uniformity of fresh concrete.

An important feature attributable to self–leveling concrete is low air content. Ordinary concrete usually has air content of 4–7%.

As adding metakaolin increases the cement-based systems viscosity, it should in turn contribute to concrete air retention. At the same time, it causes a decrease in the concrete mixtures porosity and formation of a denser concrete structure. There is research that demonstrates the insignificant effect of metakaolin on the air content of the fresh concrete [149]. During self-compaction the amount of entrapped air can be 1 – 2% [90].

Following available data, the introduction of superplasticizer can increase air entrainment up to 3%. [111]. As it is shown in Section 4.2 for fine-grained concrete, introduction of NF type superplasticizer leads to an increase in air content. However, low viscosity of self-leveling concrete causes a loss of entrained air during transportation and placing [111]. Additional reduction in air content is caused by the modification of NF superplasticizer by an air-defoaming agent (see Section 4.2).

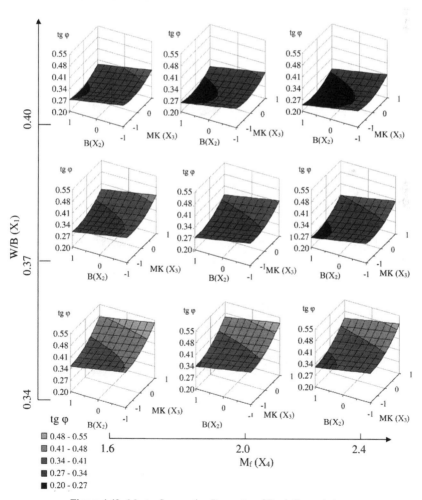

Figure 4.40. Mortar Segregation Parameter of Fresh Concrete tg φ.

Thus, there are theoretical preconditions and research results, which prove that the introduction of the composite additive of metakaolin and superplasticizer does not change the air content of mixtures. Table 4.26 represents the experimental values of air content in concrete determined by gravitational method. Based on this data, regression equation at 95% confidence probability and diagrams of air content depending on the investigated factors have been obtained:

$$V_a = 1.58 - 0.40x_1 - 0.30x_2 - 0.30x_3 - 0.40x_4 - 0.19x_1^2 + 0.09x_1^2 + 0.28x_3^2 - 0.26x_4^2$$
$$- 0.06x_1x_2 - 0.06x_3x_4 \qquad (4.40)$$
$$F_P = 4.89 < F_T = 8.77.$$

Figure 4.41 presents the influence of the studied factors on the fresh concrete air content. As can be seen, an increase in the metakaolin dosage in the binder leads to a decrease of air content (Fig. 4.41). Binder content growth

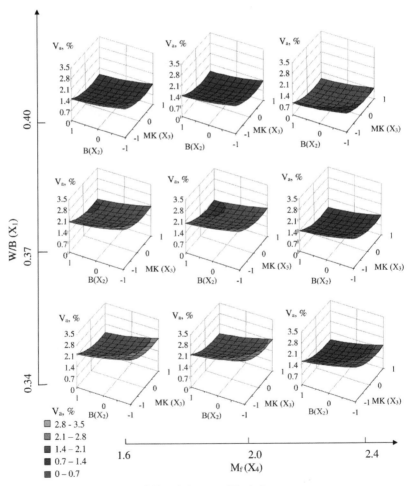

Figure 4.41. Air Content of Fresh Concrete.

leads to decreasing air content (Fig. 4.43). Regarding the effect of sand fineness modulus, it is known that the 0.15-0.6 mm fraction is the main factor influencing air entrainment, particles smaller than 0.07 mm inhibit air entrainment, and the fraction of 0.07–0.15 mm does not have a noticeable effect [150].

Taking into account the fact that the used sand with fineness modulus of 1.6 has almost a twice higher portion of 0.15–0.6 mm particles (80%) than that with fineness modulus of 2.4 (52%), it is obvious that reduction of the fineness modulus led to the increase in the air content.

Thus, adding superplasticizer and metakaolin enables it to obtain self-leveling concrete without signs of water bleeding and minor segregation values. The air content of fresh concrete is negligible and varies from 1 to 3% for all the batches. Increasing the metakaolin consumption within the considered range reduces the air content.

4.4.3 Peculiarities of Fine Aggregate Impact on Properties of Fresh Concrete

One of the main issues of concrete composition optimization is finding the ratio of fine aggregate in the total volume of aggregates. To meet the requirements of self-compacting concrete design [132], the coarse aggregate content was limited to 1250 kg/m³ (50% by solid volume) and 700 kg/m³ (40% by mortar volume). There are known methods of calculation concrete composition based on selecting the optimal portion of fine aggregate (sand) in the mixture of aggregates (sand S and crushed stone CS) [13, 88]:

$$r = \frac{S/\rho_s}{S/\rho_s + CS/\rho_{cs}} \tag{4.41}$$

where S, CS are the sand and crushed stone consumption correspondingly, kg per 1 m³ of concrete;

ρ_c, ρ_{cs} are real densities of sand and crushed stone, kg/m³.

There is an approach that the selection of this ratio is based on the rule of optimal sand content in the mixture of aggregates (r) for concrete with specified strength and workability. According to this approach, given cement content, fresh concrete has highest workability or accordingly, the lowest water consumption only at a certain sand dosage [151]. Selection of optimal sand content (r_o) can also be influenced by the need to provide the required amount of entrained air.

According to [13], there are three options for optimization of parameter r:

1. $r \to r_o$ at $\eta_{f.c.} \to$ min ($W \to$ min), where $\eta_{f.c}$ is the concrete mixture viscosity, Pa·sec;

   W is total water content of concrete, l/m³;

2. $r \to r_0'$ at $W_b \to$ min, where W_b is bleeded water content, l/m³;

3. $r \rightarrow r_0''$ at $V_a \rightarrow V_a'$, where V_a, V_a' are the actual and theoretical volumes of entrained air, % correspondingly.

As a rule, $r \leq r_0' \leq r_0''$. For self-leveling concrete, as it is specified in subsection 4.4.2, the air content is insignificant. There is a need in minimum water content in concrete to achieve high workability and eliminate water bleeding to optimize the value of r, especially to reach a higher strength of concrete. Therefore, the search for an optimal portion of sand should be initiated with calculation of r_0, from the point of the lowest viscosity of fresh concrete (minimum water content), and then verification of water bleeding possibility.

The following experimentally based equations [94] are suggested for finding r_0 and r_0'. Thus, r_0 value can be calculated as

$$r_0 = K_{C/W} r_0^{\delta} \tag{4.42}$$

where $K_{W/C}$ is coefficient, which takes into account the fresh concrete viscosity at cement-water ratio (C/W) change and constant thickness of cement paste film at aggregate grains $\delta = const;$

r_0^{δ} is the optimal value of r from the point of achieving maximum thickness of cement paste film.

To estimate water retaining ability WR of fresh concrete the following formula was suggested [105]:

$$WR = 1.35 \cdot K_{NC} \cdot C + W_s \cdot S + 0.07 \cdot s_{cs} \tag{4.43}$$

where K_{NC} is coefficient of cement paste normal consistency;

C, S are cement and sand consumption, kg/m³;

W_s is water consumption of sand, l/m³;

s_{cs} is specific surface of crushed stone, m²/m³.

According to [94], the equation has the following form:

$$WR = 1.35 \cdot K_{NC} \cdot C + K_w \cdot W_s^0 \cdot r \cdot (1 - V_{c.p}) + 0.07 \cdot s_{CS} \tag{4.44}$$

Therefore, at W = WR equation for r_0' looks as follows:

$$r_0' = \frac{W - 1.35 K_{NC} \ c - 0.07 s_{CS}}{W_s^0 (1 - V_{c.p.}) \rho_s} \tag{4.45}$$

where $V_{c.p.}$ is the cement paste volume, l/m³;

ρ_s is the sand density, kg/m³.

For self-leveling concrete, fine aggregate content must be sufficient to avoid water bleeding, which is traditionally achieved by raising r up to 40–50% [86, 88]. At the same time according to [89], self-compacting and self-leveling concrete technologies are associated with limitation of general consumption of aggregates: coarse one – less than 50% by volume of concrete solid components (it corresponds approximately to 1250 kg/m^3 for crushed stone of dense metamorphic rocks), for fine aggregate—less than 40% by cement-sand mortar volume ($\sim$ 700 kg/m^3). Therefore, the following condition should be added to optimization terms of $r : r \rightarrow r_o'''$ at $CS \rightarrow CS_{min}$. Calculations results are given in Table 4.26.

At the proportioning of fresh concrete with composite additive the value of r was determined according to graphic dependencies [88], taking into consideration binder content instead of cement content. For optimization of the accepted r values formulas given above were modified to take into consideration superplasticizer and metakaolin.

To determine r_0:

$$r_0 = K_{B/W} \cdot r_0^\delta \tag{4.46}$$

where $K_{B/W}$ is the coefficient that takes into account the change in fresh concrete viscosity and constant thickness of the binder paste film on the aggregate grains.

When calculating according to Eq. (4.46) the binder content (cement + metakaolin), m^3/m^3 was considered:

$$V_B = (W + \frac{C}{\rho_c} + \frac{MK}{\rho_{MK}})/1000 \tag{4.47}$$

To determine r_0' the following formula was used:

$$r_0' = \frac{W - 1.35 \cdot K_{NC} \cdot B - 0.07 \cdot s_{CS}}{W_s^0 \cdot (1 - V_B) \cdot \rho_s} \tag{4.48}$$

where K_{NC} is the normal consistency coefficient with SP and MK, which is calculated according to Eq (4.12), $K_{NC} = NC/100$;

B is binder consumption, kg/m^3.

Calculations, performed for concrete compositions are presented in Table 4.26. The value of r_0 was determined according to equations presented in [13].

As can be seen from the Table 4.27, for all concrete compositions $r > r_0, r > r_0'$. The values of r and r_0' at certain points of the matrix are significantly undervalued and the crushed stone consumption significantly exceeds the allowable values, which makes it difficult to obtain self-leveling concrete. Condition (3) is fulfilled for concrete of all compositions under conditions (1) and (2) (see p. 117–118). Thus, when adding metakaolin as fine mineral admixture, the proportion of sand in the mixture of aggregates should be reduced by 5–10% compared to suggested values for concrete without admixtures [88] and conditions (1) and (2) should be checked according to Eqs. (4.42) and (4.48).

Table 4.27. Optimal Sand Content for Fresh Concrete

No.	Components, kg/m³			B/W	W^0_s %	WRl/m³	K_{NC}	K_{CW}	Exp. values		Calculated values			
											$\eta_{fc} \to min$		$W_b \to min$	
	W	C	MK						r	CS, kg/m³	r_0	CS, kg/m³	r'_0	CS, kg/m³
1	220	468	82	2.50	6	226	0.25	1.29	0.345	1070	0.230	1258	0.285	1151
2	220	468	82	2.50	8	230	0.24	1.29	0.340	1078	0.221	1273	0.238	1227
3	220	523	27	2.50	6	221	0.24	1.29	0.345	1078	0.236	1258	0.301	1132
4	220	523	27	2.50	8	221	0.23	1.29	0.340	1086	0.237	1273	0.306	1124
5	180	383	67	2.50	6	186	0.22	1.29	0.375	1147	0.309	1268	0.294	1277
6	180	383	67	2.50	8	195	0.22	1.29	0.370	1156	0.306	1273	0.247	1361
7	180	428	22	2.50	6	187	0.23	1.29	0.375	1153	0.312	1269	0.279	1310
8	180	428	22	2.50	8	194	0.22	1.29	0.370	1162	0.309	1274	0.250	1362
9	187	468	82	2.94	6	208	0.22	1.34	0.345	1129	0.281	1240	0.103	1525
10	187	468	82	2.94	8	214	0.21	1.34	0.340	1138	0.274	1252	0.117	1499
11	187	523	27	2.94	6	211	0.23	1.34	0.345	1137	0.285	1241	0.074	1584
12	187	523	27	2.94	8	215	0.22	1.34	0.340	1146	0.279	1252	0.110	1523
13	153	383	67	2.94	6	177	0.21	1.34	0.375	1193	0.345	1251	0.129	1638
14	153	383	67	2.94	8	190	0.21	1.34	0.370	1202	0.343	1254	0.102	1690
15	153	428	22	2.94	6	178	0.21	1.34	0.375	1199	0.347	1252	0.125	1654
16	153	428	22	2.94	8	186	0.20	1.34	0.370	1209	0.346	1255	0.129	1646
17	200	450	50	2.50	7	206	0.23	1.29	0.360	1113	0.273	1264	0.286	1224
18	170	450	50	2.94	7	195	0.21	1.34	0.360	1166	0.315	1247	0.131	1560
19	204	495	55	2.70	7	220	0.23	1.32	0.345	1103	0.255	1255	0.178	1365
20	167	405	45	2.70	7	186	0.22	1.32	0.365	1192	0.327	1264	0.185	1506
21	185	425	75	2.70	7	200	0.22	1.32	0.360	1136	0.293	1255	0.215	1374
22	185	475	25	2.70	7	200	0.22	1.32	0.360	1143	0.297	1256	0.215	1381
23	185	450	50	2.70	6	196	0.22	1.32	0.365	1131	0.297	1252	0.226	1358
24	185	450	50	2.70	8	202	0.21	1.32	0.355	1148	0.292	1260	0.207	1391

4.5 Properties of High-strength Concrete Containing Metakaolin

4.5.1 Compressive Strength of Concrete and Kinetics of Hardening

Concrete Compressive Strength

High-strength concrete is considered as concrete with a characteristic cylinder compressive strength higher than 6000 psi (42 MPa) or greater according to ACI Committee 211 [91, 152]. According to other references high strength concrete has cube strength greater C50/60 up to 105 MPa [153]. Such strength can be achieved by fulfilling a number of requirements [91]:

1) high-grades cements and its content;

2) high-quality aggregates;

3) extremely low water-cement (water/binder) ratio due to high-range-water reducers application;

4) high allowable binder content (cement + active mineral admixtures);

5) thorough mixing and compaction of concrete mixture;

6) creation of the most favorable conditions for concrete hardening.

Manufacturing self-leveling high-strength concrete is impossible without the use of special admixtures. In order to assess the effect of metakaolin on the strength of self-leveling high-strength concrete and the kinetics of hardening, an experimental program was implemented according to a four-factor three-level plan B4 as shown in subsection 4.4, the planning conditions of the experiment are given in Table 4.28. There the compressive strength of concrete was determined at 3, 7 and 28 days using 10 x 10 x 10 cm cubic specimens.

Table 4.28. Terms of Experimental Planning.

Factors designation		Variation level		
Designation	Name of factor	Bottom (−1)	Mean (0)	Upper (+1)
X_1	Water-binder ratio (W/B)	0.34	0.37	0.40
X_2	Binder content (B), kg/m³	450	500	550
X_3	Metakaolin portion (MK), % by binder content	5	10	15
X_4	Fineness modulus of sand (M_f)	1.6	2.0	2.4

The planning matrix and results of the experiments are given in Table 4.29. Based on experimental data, adequate at 95% probability regression equations were obtained. The equations characterize the influence of the studied factors on concrete compressive strength at normal curing at 3 days:

$$f_c^3 = 41.6 - 8.0x_1 + 2.7x_2 + 0.9x_3 + 2.7x_4 - 1.0x_1^2 - 1.6x_2^2 - 1.5x_3^2$$
$$- 1.4x_4^2 + 1.5x_1x_2 + 1.0x_1x_3 + 1.2x_2x_3 \qquad (4.49)$$
$$F_D = 6.35 < F_T = 8.76;$$

at 7 days:

$$f_c^7 = 56.3 - 9.6x_1 + 3.0x_2 + 1.5x_3 + 3.0x_4 - 1.8x_1^2 - 1.5x_2^2 - 2.8x_3^2$$
$$- 1.3x_4^2 - 1.4x_1x_2 + 1.1x_1x_3 + 1.6x_2x_3 \qquad (4.50)$$
$$F_D = 5.93 < F_T = 8.76;$$

at 28 days:

$$f_c^{28} = 76.5 - 9.9x_1 + 3.5x_2 + 2.0x_3 + 2.8x_4 - 3.0x_1^2 - 1.0x_2^2 - 4.2x_3^2$$
$$- 1.3x_4^2 - 0.9x_1x_2 + 1.5x_1x_3 + 2.7x_2x_3 \qquad (4.51)$$
$$F_D = 1.49 < F_T = 8.76.$$

Table 4.29. Planned Experiment Matrix and Experimental Values of Concrete Specimens Compressive Strength.

No.	Natural values of factors				Compressive strength, MPa		
	X_1 (W/B)	X_2 (Binder)	X_3 (MK)	X_4 (M$_f$)	f_c^3	f_c^7	f_c^{28}
1	0.40	550	15	2.4	36.0	47.7	69.2
2	0.40	550	15	1.6	30.4	42.4	63.3
3	0.40	550	5	2.4	30.5	41.2	56.5
4	0.40	550	5	1.6	26.1	35.5	52.2
5	0.40	450	15	2.4	28.4	40.8	59.1
6	0.40	450	15	1.6	24.4	35.0	53.0
7	0.40	450	5	2.4	28.4	39.8	56.9
8	0.40	450	5	1.6	24.9	35.2	51.8
9	0.34	550	15	2.4	52.9	70.5	88.2
10	0.34	550	15	1.6	43.4	63.9	81.9
11	0.34	550	5	2.4	49.9	66.8	80.4
12	0.34	550	5	1.6	45.2	62.8	76.6
13	0.34	450	15	2,4	42.6	56.1	74.7
14	0.34	450	15	1,6	36.1	49.5	66.8
15	0.34	450	5	2,4	46.0	60.9	78.0
16	0.34	450	5	1,6	40.7	53.0	72.6
17	0.40	500	10	2	33.2	44.1	63.9
18	0.34	500	10	2	48.4	65.0	83.4
19	0.37	550	10	2	42.8	59.4	79.2
20	0.37	450	10	2	37.4	50.4	71.9
21	0.37	500	15	2	40.9	54.3	74.4
22	0.37	500	5	2	39.5	52.9	70.5
23	0.37	500	10	2.4	42.7	59.1	77.7
24	0.37	500	10	1.6	37.9	52.5	72.9

Figures 4.42–4.44 present diagrams of concrete compressive strength dependences on the variation factors values at 3, 7 and 28 days of hardening (Table 4.27). As can be seen from the figures and from Eqs. (4.49) – (4.50), at the early hardening period (3 and 7 days) the concrete strength slightly depends on metakaolin consumption under other constant conditions (Fig. 4.42,4.43). At the late curing period (28 days) dependence of concrete compressive strength on metakaolin consumption has extremum, corresponding to maximum strength value (Fig. 4.46). As the value of W/B increases, the optimal metakaolin content also grows. This phenomenon can be explained by a corresponding increase in water consumption required for hydration reactions. Conversely, under limited hydration conditions (at low W/B values and high metakaolin consumption), the amount of water in the mixture may not be sufficient for completion of the hydration process. Due to the pozzolanic reaction of metakaolin with $Ca(OH)_2$

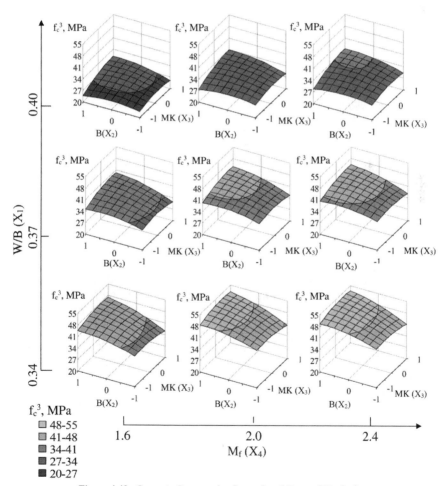

Figure 4.42. Concrete Compressive Strength at 3 Days of Hardening.

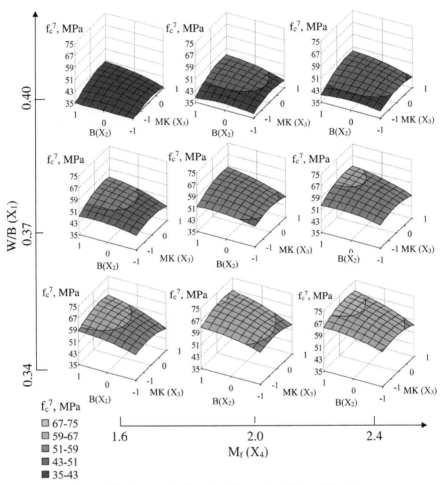

Figure 4.43. Compressive Strength of Concrete at 7 Days of Hardening.

and formation of the fine-crystalline structure, concrete compressive strength increase is observed. However, at high consumption of metakaolin (15% by binder weight), metakaolin particles, which don't participate in reactions, block the growth of hydration products, which leads to strength decreasing.

Increase in the total binder content leads to a strength increase, especially at the late hardening period (28 days). Increase in metakaolin percentage in the binder also has an extreme effect on strength. The value of the extremum increases with increasing the binder content. Thus, at a binder content of 450 kg/m³, the maximum strength value is achieved by replacing 10% of cement with metakaolin, and at a consumption of 550 kg/m³– by replacing 13%. This is due to the fact that at low binder content, replacing a significant portion of cement with metakaolin causes a deficiency of cement required for the $Ca(OH)_2$ formation, which is able to interact with metakaolin. At minimum content of metakaolin

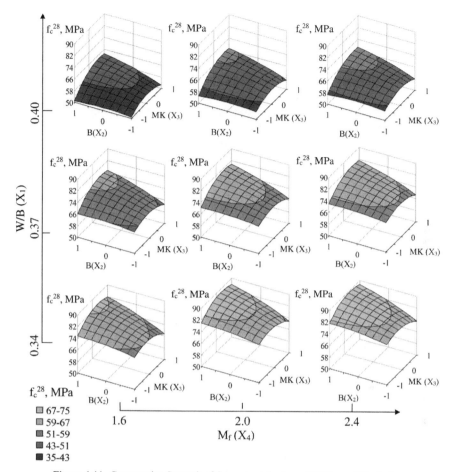

Figure 4.44. Compressive Strength of Concrete at the Age of 28 Days of Hardening.

(5%), increase in binder consumption has a minor effect on strength variation. With transition of binder content from the lower to the upper variation level, this difference becomes more noticeable. Almost equal concrete strength at binder consumption within 500–550 kg/m³, and metakaolin content of 5% should also be noted.

This phenomenon is due to the fact that increasing of cement consumption leads to growing of the cement paste film thickness on the aggregate grains. Strong crystalline structure appears in places of close contact between aggregate grains and cement hydration products [154]. However, as the distance from the point of contact increases, the adhesive strength decreases.

Under other stable conditions, the maximum values of concrete strength are observed when replacing 11–12% of the binder with metakaolin at any size of used sand (W/B = 0.37; Binder = 500 kg/m³).

It is known that decrease in sand fineness modulus, i.e., in fact, increasing in the fine aggregate surface facilitates higher adhesion with cement at a constant

W/B and leads to increase in concrete strength [155]. The strength of the contact zone especially grows with introduction of active mineral admixtures [156, 157]. However, under the current terms of the experiment reduction of sand fineness modulus leads to an increase in superplasticizer content, required to compensate the water demand growth and save equal workablity. The superplasticizer dosage above the recommended value causes reduction in compressive strength. The higher content of dusty particles in fine sand also adversely affects the strength.

Based on the experimental data, given above, it can be concluded that the optimal content of metakaolin in terms of achieving maximum strength is 10–13% by binder weight, depending on other studied factors.

Compressive strength of concrete with PC type SP (Table 4.30) was obtained according to a three-factor plan based on sand with fineness modulus of 1.6 (see concrete compositions in Table 4.24).

The regression equation is given below:

$$f_c^{28} = 77.0 - 9.4x_1 + 2.6x_2 + 3.9x_3 - 1.4x_2^2 - 1.8x_3^2 + 0.8x_1x_3 + 1.6x_2x_3 \quad (4.52)$$
$$F_D = 6.71 < F_T = 19.37.$$

Concrete compressive strength has a linear dependence on W/B ratio change at constant values of other factors (Fig. 4.45). As well as for concrete containing NF type SP, W/B ratio has the most significant effect on compressive strength

Table 4.30. Matrix of Planning and Results of Testing Compressive Strength (PC type SP).

Test No.	Matrix of planning			Compressive strength at the age of 28 days, MPa
	W/B X_1	Binder, kg/m³, X_2	Metakaolin, %, X_3	
1	0.4	550	15	74.0
2	0.4	550	5	61.5
3	0.4	450	15	65.5
4	0.4	450	5	58.0
5	0.34	550	15	91.7
6	0.34	550	5	81.0
7	0.34	450	15	82.0
8	0.34	450	5	79.4
9	0.4	500	10	68.0
10	0.34	500	10	87.2
11	0.37	550	10	77.1
12	0.37	450	10	74.3
13	0.37	500	15	78.0
14	0.37	500	5	72.6
15	0.37	500	10	76.0
16	0.37	500	10	77.0
17	0.37	500	10	76.5

compared to all other factors. The W/B effects are similar for both superplasticizers and they don't depend on the SP type. Following available data, superplasticizers based on PC can be ineffective for Portland cement—metakaolin blends [158]. Whereas another research confirms the high effect of superplasticizers based on polycarboxylate ethers on water reduction and compressive strength of concrete containing metakaolin [159]. W/B change from bottom to upper level yields an increase of about 9.4 MPa in the concrete compressive strength. Increasing the binder content and subsequent rise in metakaolin portion in it leads to a positive effect on compressive strength. The linear effect of binder content (X_2) is slightly lower than effect of metakaolin content (X_3). Negative values of quadratic terms for these factors (X_2 and X_3) are also significant. Therefore, these factors have extreme influence on concrete compressive strength. Combined effect is observed at interaction of factors X_1 - X_3 and X_2 - X_3. It proves that optimal metakaolin content increases at W/B and binder content rising. This phenomenon as well as for NF type SP concrete can be explained by the respective increasing in water content, required for hydration reaction, whereas in "tight" hydration conditions (at low W/B values and high metakaolin content) there is a lack of water for the normal hydration reaction [160]. Due to the reaction of alumina and silica, contained in metakaolin with $Ca(OH)_2$, and fine-crystalline dense structure formation, an increase in concrete strength is observed. However, as well as for NF-based SP concrete at upper level of X_3 (15% by binder mass) metakaolin particles, which don't take part in reaction, block the new formations growth and yield concrete strength reduction. Therefore, for SP of PC type-based concrete when binder content is 450 kg/m³ at W/B = 0.34 an optimal metakaolin content is about 12% and at W/B = 0.4 it is about 15% (see Fig. 4.45).

It should be mentioned that at the bottom level of metakaolin portion in binder an insignificant concrete strength growth (about 2–3 MPa) is observed, whereas at the upper metakaolin content level the difference becomes more significant (about 10 MPa). Concrete strength for binder content within 500–550 kg/m³ does not change significantly at low metakaolin portion and does not depend on the SP type used.

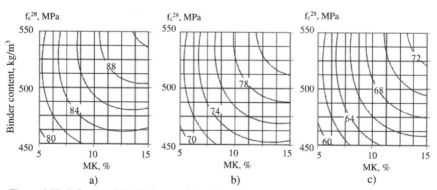

Figure 4.45. Influence of Binder Content (x_2) and Metakaolin Portion in Binder (x_3) on Compressive Strength of SP PC Type-based Concrete at 28 Days: a) W/B = 0.34; b) W/B = 0.37; c) W/B = 0.40.

Kinetics of High-strength Concrete Hardening

To determine compressive strength of concrete at each point of the four-factors plan (see Table 4.28) for concrete based on SP NF type at any time $n \le 28$ days, a known equation can be used [1]:

$$f_c^n = f_c^k + (f_c^{28} - f_c^k) \frac{lg\,n - lg\,k}{lg\,28 - lg\,k} \qquad (4.53)$$

To find the adequacy of Eq. (4.53) was calculated, taking $k = 3$ days. Calculated values and mean experimental values of are given in Table 4.31.

Table 4.31. Calculated and Experimental Values f_c^7.

Batch No.	1	2	3	4	5	6	7	8	9	10	11	12
f_{exp}^7	47.7	42.4	41.2	35.5	40.8	35.0	39.8	35.2	70.5	63.9	66.8	62.8
f_{calc}^7	48.6	42.9	40.4	36.0	40.0	35.2	39.2	35.1	66.3	58.0	61.5	57.1
Batch No.	13	14	15	16	17	18	19	20	21	22	23	24
f_{exp}^7	56.1	49.5	60.9	53.0	44.1	65.0	59.4	50.4	54.3	52.9	59.1	52.5
f_{calc}^7	54.8	47.8	58.2	52.8	44.9	61.6	56.6	50.5	53.6	51.2	56.0	51.2

The equation is adequate as $F_d = 7.68 \le F_t = 8.65$.

To estimate the strength kinetics for each batch at the hardening rate at the age of 3 and 28 days was determined, which can be expressed as follows:

$$V_n = \frac{df_c^n}{dn} = \frac{(f_c^{28} - f_c^k)}{(lg\,28 - lg\,k)} \cdot \frac{0.4343}{n} \qquad (4.54)$$

Figure 4.46 presents kinetics of concrete hardening by compressive strength values at 28 days. As can be seen, the hardening kinetics significantly depends on concrete composition.

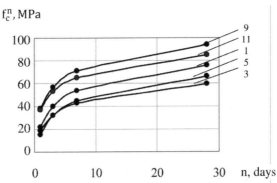

Figure 4.46. Kinetics of Concrete Hardening (# according to Table 4.31).
1 – W/B=0.40; B=550 kg/m³; MK=15%; M_f=2.4; 3 – W/B =0.40; B=550 kg/m³; MK=5%; M_f=2.4; 5 – W/B =0.40; B =450 kg/m³; MK =15%; M_f=2.4; 9 – W/B =0.34; B =550 kg/m³; MK =15%; M_f=2.4; 11 – W/B =0.40; B =550 kg/m³; MK =5%; M_f=2.4.

To perform a more detailed analysis of the studied factors influence on hardening kinetics at the early and late curing periods, obtained on the basis of calculations by Eq. (4.54), regression equations of curing rate r are as following: at 3 days:

$$r_3 = 5.21 - 0.28x_1 + 0.12x_2 + 0.26x_3 - 0.31x_1^2 - 0.40x_3^2 - 0.07x_1x_3$$
$$+ 0.21x_2x_3 \tag{4.55}$$

at 28 days:

$$r_{28} = 0.55 - 0.03x_1 + 0.01x_2 + 0.03x_3 - 0.03x_1^2 - 0.04x_3^2 + 0.01x_1x_3$$
$$+ 0.02x_2x_3 \tag{4.56}$$

Figure 4.47 shows a dependence of hardening rate on the considered factors. As can be seen, with a decrease of W/B from 0.37 to 0.34, the hardening rate does not change significantly. It can be explained by the large amount of NF type superplasticizer, which retards down the initial structure formation processes. It should be noted that the change in the metakaolin content and W/B have the greatest influence on the hardening rate (Eqs. (4.55) and (4.56)). According to the known data [1], W/C has a more significant impact on concrete strength at 1–3 days than at the late hardening stages, and the lower W/C is, the greater rate of hardening is. Increasing the binder content accelerates hardening rate. It can be explained by the fact that in conditions of limited water content, the structure formation processes and formation of contacts between hydration products occur faster. An increase in the metakaolin dosage in the binder has an extreme effect on hardening rate. As binder content grows, the optimal content of metakaolin in it increases (Fig. 4.48). As can be seen from the regression equations (4.55) and (4.56), the hardening rate at 28-th day is an order of magnitude lower than that at the 3rd day, as evidenced by the regression coefficients values of the respective equations.

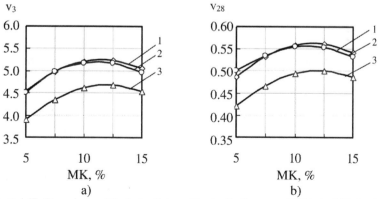

Figure 4.47. Dependence of Hardening Rate on Metakaolin Content and W/B: 1 – 0.37; 2 – 0.34; 3 – 0.40; (B=500 kg/m³, M_f=2.0) at: a) 3 days; b) 28 days.

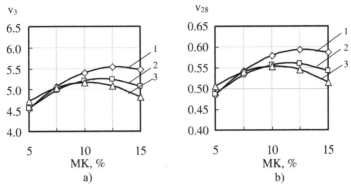

Figure 4.48. Dependence of Hardening Rate on Metakaolin Dosage and Binder Content:
1 – 550 kg/m³; 2 – 500 kg/m³; 3 – 450 kg/m³; (W/B=0.37, M_f=2.0) at: a) 3 days; b) 28 days.

Thus, kinetics of concrete hardening is affected by its composition. Increasing the metakaolin content in binder by the amount of 10–13% by weight leads to an increase in the concrete hardening rate. At a higher metakaolin content, due to the blocking of structure formation processes by metakaolin particles that do not participate in reaction, the hardening rate decreases. Acceleration of early stage hardening at increasing metakaolin content can be explained by additional compaction of concrete and the formation of dispersed hydration products with higher strength.

4.5.2 *Efficiency Factor of Metakaolin Application for Self-leveling High-strength Concrete*

Efficiency Criteria for Mineral Admixtures: Review and Selection

Some aspects of determining mineral admixtures efficiency were described in subsection 4.2.3. Methods for determining mineral admixtures efficiency can be classified by the estimation object, depending on the effects on a certain parameter. The chemical effect is measured by $Ca(OH)_2$ absorption [143], the physical-mechanical effect is mostly determined by the compressive strength increase [39, 161], the physical-chemical effect is based on calorimetry of hydrated cement and concrete measurements [93, 95], the design effect results in concrete design approach change [4, 162, 163].

There is research, that considers the efficiency factor (*k*-value) as a part of supplementary cementing materials (SCM), equivalent to Portland cement, which provides the same concrete properties (e.g. compressive strength, chloride permeability, etc.) as for the concrete without SCM [47, 163]. This effect reflects the influence of mineral admixture on concrete composition proportioning. Numerical values of efficiency factors for different mineral admixtures were studied and are reported in the literature [47, 163]. However, the influence of the superplasticizer's type and its content change due to mineral admixture introduction was not investigated. A possible increase in concrete strength due to mineral admixtures addition as well as selecting the admixtures' dosage were also out of the research scope.

Combined effects reflect the influence of mineral admixtures on concrete strength, slump, frost resistance, durability, etc., as well as on concrete manufacturing particularities like changing the concrete composition, especially decreasing the part of cement and varying the chemical admixtures [94, 95, 164]. The most distinctive feature of the approach [94] is that different effects on the mortars compressive strength are separated. Three main effects of mineral admixture on mortars strength have been quantified as follows [95]:

- effect on cement dilution (replacement) by mineral admixture;
- physical effect due to filler effect and heterogeneous nucleation;
- chemical effect due to pozzolanic reaction.

Following the literature, both inert (limestone filler, crushed quartz) and pozzolanic admixtures (fly ash) were used. A more detailed review of available methods for estimating the mineral admixtures efficiency is given in Table 4.32. There is analyzed the combined effect of metakaolin as a mineral admixture based on available experimental results.

It is common to pay more attention to reducing the cement content or changing the mechanical and performance properties of concrete at estimating the mineral admixtures efficiency [47, 143, 165–167]. Important parameters, affecting the modern mineral admixtures efficiency, include not just cement content change, but also higher concrete performance characteristics, like its strength properties, permeability resistance to aggressive environments, etc. [39, 91, 161]. This issue is especially important when high-reactive pozzolanic mineral admixtures like silica fume and metakaolin are used. The observed increase in concrete compressive strength has often appeared to be the easiest for implementation and interpretation of efficiency factor attributable to mineral admixtures [39, 47, 162]. For proper design of reinforced concrete structures to withstand strong earthquakes, wind and other dynamic loadings ductility should be also considered [168].

However, due to high water consumption, most of such reactive admixtures become effective only in combination with high dosages of superplasticizers [90, 91, 166, 169]. A meaningful efficiency factor determination has to consider the concrete strength increase, obtained by increasing the SP dosage and reduced cement content.

Analysis of the estimated mineral admixtures' effects shows that an efficiency factor, representing the combined effect of cement and SP contents on concrete strength, is the most adoptable to practical concrete manufacturing conditions [94]. This factor reveals the combination of design parameters and the mineral admixture physic-mechanical effects on the produced concrete properties. The described below efficiency factor calculation method [94, 144, 170] can be applied both for pozzolanic and cementitious mineral admixtures.

Table 4.32. Parameters of Mineral Additive Effectiveness (CBM).

Effect	Investigated parameter	Advantages	Reference
1	2	3	4
Chemical effect	Content of Ca(OH)$_2$ absorbed by pozzolanic admixture	One of the most common and oldest parameters, a lot of data availability for different admixtures for comparing. Dependent only on cement composition and independent of concrete components and manufacturing technology.	130
Physical-mechanical effect	Compressive strength, strength activity index	Simplicity and clearness of method and results obtained, effective for comparing different mineral admixtures, suitable in field conditions.	23, 149
Physical-chemical effect	Heat release rate and other calorimetric data	Determining separate effects of various chemical and mineral admixtures on cement and concrete hardening and their mutual compatibility. Applicable for early hardening.	68, 77
Design effect	Reduced cement-water ratio,coefficient of cementitious efficiency, portion of cement replaced	Revealing cement saving, suitable for by-product admixtures such as fly ash, blast-furnace slag, silica fume, suitable for calculating improvement in sustainability of cement-based concrete and for concrete mixtures design with limited or small amount of cement.	91, 150, 151
Combined effect	Cement dilution, heterogeneous nucleation,compressive strength	Revealing the effect on concrete proportioning, short-term and long-term hardening processes, separation of effects caused by different factors.	80
	Water requirement, compactness, occluded air volume, workability, compressive strength	Combination of chemical, physical-chemical, microstructural and other effects of different admixtures by applying classical models, generalized for various types of cements and a significant number of fine and ultra-fine admixtures.	152
	Portion of replaced cement, compressive strength, superplasticizer content change	Shows cement and superplasticizer content change at simultaneous strength increase, suitable for practical use in concrete design.	78

A method for Determining the Mineral Admixture Efficiency Factor

Determining the efficiency factor, as described in [94] has considerable practical interest. Research data on this subject indicates that such factors can be a good criterion for assessing pozzolanic efficiency. The following equation can be used for determining the factor [94]:

$$F_c = R_i / \sqrt{[C_i + F_c(S_i - S_c)] \cdot 100} \qquad (4.57)$$

where R_i is a ratio between the strength of concrete with mineral admixture and SP to that with SP only, % for the same water-binder ratio and workability;

C_i is the relation between the cement content in the specimens with mineral admixture and SP to that with SP additive, %;

F_c is the cement content reduction factor due to the SP addition.

F_c can be determined as follows:

$$F_c = (C_o - C_c)100/(C_o S_c) \qquad (4.58)$$

where C_o, C_c are the cement contents in the control specimens without and with addition of SP respectively, kg/m³ for equal water binder ratio, workability and compressive strength class;

S_c is the SP dosage in the control specimens, %;

S_i is the SP dosage, required for obtaining equal workability for mixtures with mineral admixture and with SP, %.

The mineral admixtures are divided by their efficiency factor to high-effective mineral admixture ($F_e \geq 1.2$) and low-effective mineral admixture ($F_e < 1.2$). An important positive peculiarity of this efficiency factor is separating the effect of SP and mineral admixture on concrete strength. Applying Eq. (4.57), the values of F_e for such mineral admixtures as silica fume, tripoli, and fly ash have been determined [94]. This method can be applied for estimation metakaolin efficiency in cement-based concrete in combination with NF and PC type superplasticizers.

Equation (4.57) is limited by a comparatively narrow data range with strict limitations obtained for specimens of equal water-binder ratio and workability. Therefore, the data presented in [94] are related to low-slump concrete (S2 class) with a comparatively high water-binder ratio W/B = 0.4 which does not correspond to the self-leveling high-strength concrete requirements.

Taking into consideration the high workability of fresh concrete (S4–S5 class) we have modified the Eq. 4.57 by introducing a slump change coefficient, F_S %:

$$F_e = R_i / \sqrt{[C_i + F_s \cdot F_c(S_i - S_c)] \cdot 100} \qquad (4.59)$$

where

$$F_S = (S_o - S_c) \cdot 100/S_o \qquad (4.60)$$

where S_o, S_c are slumps of the control specimens without and with SP additive respectively.

Efficiency Factor for Metakaolin

A three-level and three-factor plan, similar to a D–optimal one was used in the experimental work as it was described for efficiency factor of metakaolin. After the specimens were prepared and tested the experimental data was used to calculate the efficiency factor of metakaolin in SP NF and SP PC type concrete.

For calculation of the efficiency factor additional experiments were carried out for concrete control compositions given in Table 4.33. The experimental data were used to calculate the efficiency factor of metakaolin in SP NF based concrete (Table 4.34) and SP PC based concrete (Table 4.35).

Table 4.33. C40/50 Class Concrete Compositions (With and Without Superplasticizer) used in Tested Control Specimens.

No.	W/C	Components content, kg per m³ of concrete						Sl, cm	f_c, MPa
		C	W	S	CS	SP NF (for 1st set of exp.)	SPPC (for 2nd set of exp.)		
1C	0.42	628	264	431	1016	-	-	16	52.0
2C	0.40	450	180	662	1165	5.18	-	22	53.0
3C	0.40	500	200	601	1129	3.00	-	22	55.0
4C	0.40	550	220	543	1090	1.54	-	22	57.5
5C	0.40	450	180	662	1165	-	0.72	22	55.0
6C	0.40	500	200	601	1129	-	0.70	22	57.1
7C	0.40	550	220	543	1090	-	0.66	22	59.4

Adequate experimental-statistical models for metakaolin efficiency factors were obtained as follows:

- for NF type SP based concrete:

$$F_{e1} = 1.43 - 0.22x_1 + 0.05x_2 + 0.17x_3 - 0.06x_2^2 - 0.02x_1x_2 + 0.04x_2x_3 \quad (4.61)$$
$$F_D = 5.89 < F_T = 19.37.$$

- for PC type SP based concrete:

$$F_{e2} = 1.48 - 0.21x_1 - 0.05x_2 + 0.15x_3 + 0.03x_1^2 - 0.03x_3^2 + 0.03x_1x_2 \quad (4.62)$$
$$F_D = 5.40 < F_T = 19.37.$$

Coefficients of Eqs. (4.61) and (4.62) indicate, that metakaolin, used with PC type SP, has a higher efficiency factor F_e, than with SP of NF type. It was observed that in both cases W/B ratio caused the same influence on efficiency factor. Moreover, the linear effect of W/B change on efficiency factor was most significant. Increase in metakaolin content has a positive effect on efficiency factor, whereas that of binder content growth is adverse. There are significant quadratic effects of the x_2 factor for NF type SP based concrete. Quadratic effects of x_1 and x_3 factors for PC type SP based concrete are less significant. The most significant combined effect is observed at interaction of x_1 and x_2 factors in both cases as well as at x_2 and x_3 interaction for SP NF based concrete.

Figures 4.49 and 4.50 present the efficiency factor dependence on the factors under investigation. It reveals that reduction of W/B leads to increase of F_e, as concrete strength increases significantly. At the same time graphical dependences of F_e on binder content have optimal values that vary depending on W/B and

Table 4.34. Calculation of Metakaolin Efficiency Factor in Combination with NF Type Superplasticizer.

Test No.	Data for calculation of efficiency factor											F_{el}
	R_1, MPa	R_{c1}, MPa	R_{i1}, %	C, kg/m³	C_c, kg/m³	C_i, %	C_0, kg/m³	S_{c1}, %	F_{c1}, %	F_{sl}, %	S_{l1}, %	
1	68.0	57.5	118.3	468	550	85	628	0.28	44.4	−0.38	0.75	1.35
2	53.6	57.5	93.2	523	550	85	628	0.28	44.4	−0.38	0.50	0.98
3	59.0	53.0	111.3	383	450	95	628	1.15	24.6	−0.38	1.45	1.23
4	51.8	53.0	97.7	428	450	95	628	1.15	24.6	−0.38	0.75	0.98
5	85.0	57.5	147.8	468	550	85	628	0.28	44.4	−0.38	1.65	1.87
6	76.6	57.5	133.2	523	550	85	628	0.28	44.4	−0.38	0.80	1.43
7	74.8	53.0	141.1	383	450	95	628	1.15	24.6	−0.38	2.25	1.63
8	71.0	53.0	134.0	428	450	95	628	1.15	24.6	−0.38	1.30	1.38
9	62.0	55.0	112.7	450	500	90	628	0.60	34.0	−0.38	0.85	1.21
10	80.3	55.0	146.0	450	500	90	628	0.60	34.0	−0.38	1.50	1.65
11	73.0	57.5	127.0	495	550	90	628	0.28	44.4	−0.38	0.80	1.41
12	65.9	53.0	124.3	405	450	90	628	1.15	24.6	−0.38	1.35	1.32
13	74.2	55.0	134.9	425	500	85	628	0.60	34.0	−0.38	1.60	1.59
14	65.2	55.0	118.5	475	500	95	628	0.60	34.0	−0.38	0.90	1.24
15	70.4	55.0	128.0	450	500	90	628	0.60	34.0	−0.38	1.23	1.41
16	71.0	55.0	129.1	450	500	90	628	0.60	34.0	−0.38	1.20	1.42
17	71.5	55.0	130.0	450	500	90	628	0.60	34.0	−0.38	1.17	1.43

Table 4.35. Calculation of Metakaolin Efficiency Factor in Combination with PC Type Superplasticizer.

Test No.	Data for efficiency factor calculation											F_{e2}
	R_2, MPa	R_{c2}, MPa	R_{i2}, %	C, kg/m³	C_c, kg/m³	C_i, %	C_0, %	S_{c2}, %	F_{c2}, %	F_{sl}, %	S_{i2}, %	
1	74.0	59.4	122.6	468	550	85	628	0.12	103.5	−0.38	0.25	1.39
2	61.5	59.4	103.5	523	550	85	628	0.12	103.5	−0.38	0.16	1.07
3	65.5	55.0	120.9	383	450	95	628	0.16	177.1	−0.38	0.38	1.42
4	58.0	55.0	105.5	428	450	95	628	0.16	177.1	−0.38	0.26	1.12
5	91.7	59.4	149.8	468	550	85	628	0.12	103.5	−0.38	0.32	1.76
6	81.0	59.4	136.4	523	550	85	628	0.12	103.5	−0.38	0.27	1.44
7	82.0	55.0	·149.1	383	450	95	628	0.16	177.1	−0.38	0.51	1.90
8	79.4	55.0	144.4	428	450	95	628	0.16	177.1	−0.38	0.39	1.62
9	68.0	57.1	120.8	450	500	90	628	0.14	145.6	−0.38	0.23	1.29
10	87.2	57.1	152.7	450	500	90	628	0.14	145.6	−0.38	0.36	1.73
11	77.1	59.4	129.8	495	550	90	628	0.12	103.5	−0.38	0.24	1.41
12	74.3	55.0	135.1	405	450	90	628	0.16	177.1	−0.38	0.34	1.53
13	78.0	57.1	136.6	425	500	85	628	0.14	145.6	−0.38	0.31	1.57
14	72.6	57.1	127.1	475	500	95	628	0.14	145.6	−0.38	0.19	1.32
15	76.0	57.1	133.1	450	500	90	628	0.14	145.6	−0.38	0.25	1.45
16	77.0	57.1	134.9	450	500	90	628	0.14	145.6	−0.38	0.26	1.48
17	76.5	57.1	134.0	450	500	90	628	0.14	145.6	−0.38	0.27	1.47

metakaolin portion due to the combined effects (interaction between factors x_1 and x_3). Thus, the higher the W/B ratio is, the lower optimal binder content under constant metakaolin portion is. For example, when metakaolin content is equal to 10% (i.e., $x_3 = 1$) at W/B change ratio from bottom to upper level the optimal binder content varies from 510 to 530 kg/m³. Increase of metakaolin portion in binder causes subsequent growth in optimal binder content, e.g., at constant W/B value, equal to 0.37, an increase in metakaolin portion in binder from 5 to 15% leads to subsequent optimal binder content growth from 500 to 530 kg/m³.

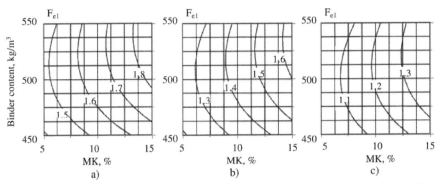

Figure 4.49. Influence of Binder Content (x_2) and Metakaolin Portion in Binder (x_3) on Metakaolin Efficiency Factor in NF Type SP Based Concrete: a) W/B = 0.34; b) W/B = 0.37; c) W/B = 0.40.

For the SP₂ based concrete the influence of binder content on efficiency factor is almost linear. So, the metakaolin efficiency at presence of SP₂ is less sensitive to the factors' combinations. In both cases the reduction efficiency factor range was observed at a higher W/B ratio. It reveals that at low W/B and high strength range it is more sensitive to factors variability.

Following Figs. 4.49 and 4.50, metakaolin is considered to be a high-effective mineral admixture in most of the experimental area points. The areas of F_e values beyond the 1.2 level curve are present at diagrams for both types of SP when W/B = 0.4. The area of low metakaolin efficiency can be described by

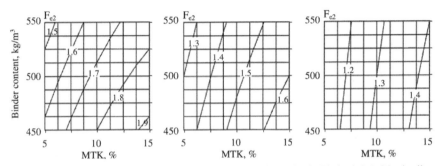

Figure 4.50. Influence of Binder Content (x_2) and Metakaolin Portion in Binder (x_3) on Metakaolin Efficiency Factor in PC Type SP Based Concrete: a) W/B = 0.34; b) W/B = 0.37; c) W/B = 0.40.

the following inequalities derived from Eqs. (4.61) and (4.62), as $F_e < 1.2$ and $x_1 = 1$ (i.e., W/B = 0.4) and the factors variation intervals can be obtained from the diagrams.

For SP NF type based concrete:

$$0.03x_2 - 0.06x_2^2 + 0.17x_3 + 0.04x_2x_3 + 0.01 > 0; x_2[-1;1], x_3[-1;0.6)$$

For SP PC type based concrete:

$$-0.02x_2 + 0.15x_3 - 0.03x_3^2 + 0.1 > 0, x_2[-1;1], x_3[-1;-0.5).$$

Using the diagrams, according to the output parameter value and other known factors it is possible to obtain the unknown factor in the arrow direction on the graph. The regression equations and graphs can be applied in concrete design for solving technological problems.

4.5.3 Crack Resistance Parameters of Concrete

Crack resistance of concrete characterizes its ability to resist crack formation caused by the complex action of both external (mechanical forces) and internal factors (shrinkage, temperature deformation). Among the properties of concrete associated with its crack resistance, the most influential properties are prismatic compressive strength, tensile strength, modulus of elasticity, shrinkage, deformability (ultimate tensile strength in axial tension). Concrete stresses, developing both under direct mechanical action and as a result of shrinkage and thermal deformations, may cause cracks. In general, cracks in concrete occur when the tensile deformations, ε_{ct}, resulting from the above factors exceed the ultimate tensile strength of concrete, ε_t.

It is known that increased content of mortar in self-leveling concrete causes increased deformability and reduced crack resistance. Additionally, self-leveling concrete without additives is characterized by lower elasticity modulus values, higher Poisson's ratio, longitudinal and transverse deformations [90, 146].

In order to assess the crack resistance using known calculation methods, strength and strain properties were studied.

Strength Properties of Concrete

Based on the experimental data given in Table 4.29, concrete compositions containing NF type SP and metakaolin, as well as control composition concrete (without metakaolin) were selected. Prism compressive strength, axial tensile and bending strength, modulus of elasticity, shrinkage deformation and criteria of crack resistance have been obtained. The study was performed using 10 × 10 × 40 cm prism specimens. Compositions are given in Table 4.36. Strength and deformation characteristics are presented in Table 4.37.

There is a certain correlation between the cubic compressive strength and each of these indicators, represented by the calculation dependences [171]. It is considered that such factors such as the concrete composition, components

Table 4.36. Compositions of Concrete Batches for Determining Concrete Properties.

No.	W/C	W/B	Basic components, kg/m³				Admixtures, % Binder/kg/m³	
			Water	Cement	Sand*	CS	SP NF	MK
1	0.40	0.40	200	500	610	1105	0.50/2.5	–
2	0.42	0.40	200	475	608	1101	0.55/2.75	5/25
3	0.44	0.40	200	450	606	1097	0.75/3.75	10/50
4	0.38	0.34	170	450	634	1149	1.35/6.75	10/50

Note: * quartz sand with fineness modulus $M_f = 2.0$ was used.

quality, method of placing, concrete age have little effect on these dependences [172].

Data on prism strength f_{pc} for the studied concrete are shown in Fig. 4.51. Based on statistical processing of experimental data, a correlation between the prism strength f_{pr} and cubic strength of concrete f_c was obtained [171]:

$$f_{pc} = K_{pc} \cdot f_c, \qquad (4.63)$$

where $K_{pc} = 0.783$ (K_{pc} is prism strength coefficient).

As can be seen from the Table 4.37, the prism strength of concrete with SP at different age reaches 80–84% of the cubic strength, and when replacing 10% of cement by metakaolin, this ratio is 85–89%. Such an increase of K_{pc} values is characteristic for high-strength concrete and was reported by many authors [86, 171, 172, 173].

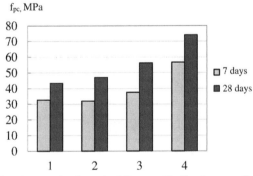

Figure 4.51. Prism Compressive Strength of Concrete (# of batches according to Table 4.36).

It is known that concrete resistance to axial tension is significantly lower than compressive strength. It is expected that the introduction of metakaolin into concrete while increasing the concrete homogeneity and improving adhesion between concrete components will lead to increasing these parameters. It confirms the experimental data (Fig. 4.53).

To calculate tensile strength f_t, Ferret has proposed the following dependence [171].

Table 4.37. Strength and Strain Parameters of Concrete.

No.	τ, days	f_c, MPa	$f_{p,c}$, MPa	K_{pc}	f_t, MPa Experimental	f_t, MPa Calculated (4.64)	f_b, MPa Experimental	f_b, MPa Calculated (4.65)	E·10⁻³MPa Experimental	E·10⁻³MPa Calculated (4.66)	E·10⁻³MPa Calculated (4.67)	E_d·10⁻³, MPa (calculated, 4.71)	f_{sf}, MPa (4.70)	μ
1	3	28.9			1.98	2.19	2.90	3.50					2.40	
	7	40.7	32.6	0.80	2.47	2.75	3.87	4.39	31.13	33.22		42.30	3.02	
	28	53.6	43.4	0.81	2.87	3.30	4.76	5.28	37.73	36.39	34.41	45.12	3.63	0.31
2	3	29.3			2.14	2.21	3.15	3.53					2.43	
	7	39.5	32.0	0.81	2.57	2.69	4.07	4.31	36.62	32.86		41.97	2.96	
	28	56.0	47.0	0.84	3.09	3.40	5.11	5.44	42.35	36.86	35.17	45.53	3.74	0.28
3	3	33.2			2.11	2.40	3.76	3.84					2.64	
	7	44.1	37.5	0.85	2.62	2.90	4.81	4.63	38.18	34.18		43.16	3.19	
	28	63.9	56.2	0.88	3.15	3.71	6.2	5.94	44.72	38.24	37.57	46.70	4.08	0.25
4	3	48.4			2.75	3.08	4.85	4.93					3.39	
	7	65.0	56.6	0.87	3.06	3.75	5.96	6.00	40.15	38.41		46.85	4.13	
	28	83.4	74.2	0.89	3.73	4.43	6.62	7.09	46.35	40.76	42.92	48.79	4.87	0.24

f$_t$, MPa

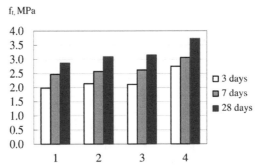

Figure 4.52. Axial Tensile Strength of Concrete (# of batches according to Table 4.36).

$$f_t = 0.05 \, (10 f_c)^{2/3} \tag{4.64}$$

obtained for low-class concrete, and later extended it to C50 concrete.

As can be seen from Table 4.37, the experimental data indicates slightly lower axial tensile strength of the concrete of these compositions than calculated by Eq. (4.64). The low tensile strength, compared to the data corresponding to the Fere formula, proves that it is still more suitable for low-strength compressive concrete. For such concrete, the ratio of tensile to compressive strength is higher than for high–strength concrete (Fig. 4.52).

For high-strength concrete compressive strength grows faster than tensile and bending strengths. According to [106], this ratio varies over time from 15–18% at 3 days to 6–7% at 180 days.

Results of concrete bending strength are shown in Fig. 4.53.

f$_b$, MPa

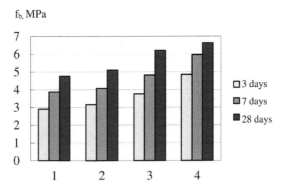

Figure 4.53. Bending Strength of Concrete (# of batches according to Table 4.36).

The following correlation between concrete bending (flexural) strength of and cubic compression strength is known [159]:

$$f_b = 0.08 \, (10 \cdot f_c)^{2/3} \tag{4.65}$$

Experimental results (Table 4.37) show an increase in the corresponding strength characteristics of concrete with a complex SP+MK additive, compared to control samples without metakaolin, but the values are lower than those calculated by Eq. (4.65).

It should be noted that for concrete specimens of equal compressive strength class (C40, compositions 1 and 2), the values of axial tensile and bending strength increase, when replacing 5% cement by metakaolin. It corresponds to known data for self-leveling high-strength concrete [174].

Deformation Parameters

Concrete deformations were determined by testing 10x10x40 cm prisms at 7 and 28 days to the loading level of ~ 0.8 R_p (R_p is destructive force). The initial modulus of elasticity and Poisson's ratio were determined at $\sigma = 0.3$ f_p. Figure 4.54 shows the dependence of concrete longitudinal and transverse deformations on the load level at 28 days.

As can be seen from Fig. 4.54, concrete of control composition has slightly higher deformations than concrete with addition of SP and MK. There is almost no increase in the relative longitudinal and transverse deformations of rapid-rise creep within the measured deformations, which is characteristic for high-strength concrete [86]. The greater the concrete strength is, the more linearized the σ-ε graphical dependence is (curves 2 and 4, Fig. 4.56).

Such deformation features, characteristic for high-strength concrete (low values or absence of plastic deformations on the ascending branch of the diagram) cause the main disadvantage of high-strength concrete—its fragile nature of compression fracture.

Increasing the metakaolin dosage in binder leads a growth in modulus of elasticity and a decrease in Poisson's ratio (Table 4.37). The approximate

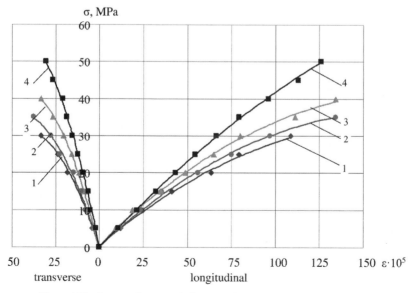

Figure 4.54. Concrete Stress-strain Curve at 28 Days (according to Table 4.36).

modulus of elasticity value of concrete can be calculated by the known formula [171]:

$$E = 52000 \cdot f_c/(23 + f_c) \qquad (4.66)$$

The values of the modulus of elasticity, calculated by this equation, have a satisfactory convergence for concrete with actual values of $E \leq 40 \cdot 10^3$ MPa.

The ACI code suggests the following equation:

$$E = 4700 \sqrt{f_c} \qquad (4.67)$$

where f_c is the concrete compressive strength at 28 days. The calculation results by this equation are given in Table 4.36. The values are lower than the experimental ones.

Long-term deformations of concrete were also investigated: drying shrinkage was considered as one that causes the largest deformations of concrete (see subsection 4.3.3). It is explained by the action of capillary forces that occur in cement paste during water evaporation from capillaries, removal of intercrystalline water and adsorption-bound from tobermorite gel [116]. Self-leveling concrete, as having the excessive free water, is characterized by higher shrinkage than plastic mixtures [90]. In addition, it is known that significant content of binder, characteristic of cast high-strength concrete, increases its contraction shrinkage [111, 151]. Shrinkage deformations cause internal stresses in concrete, which can lead to cracking in the contact zone and cause low frost and water resistances.

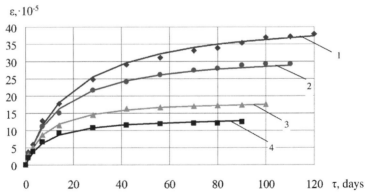

Figure 4.55. Shrinkage Deformations of Concrete (batches numbers correspond to Table 4.36).

As it is shown by the above dependences, shrinkage deformations of concrete with SP and MK additives (batches 2–4) stabilize at the age of 60–80 days, while for superplactisized concrete—at 100–120 days (batch 1) (Fig. 4.57). At the end of the experimental period, the shrinkage deformations of concrete samples with metakaolin were 2–3 times lower than without metakaolin, and a significant part of shrinkage deformations in specimens containing metakaolin ($\sim 70\%$) appeared in the first 28 days of hardening. This effect can be explained by high-water retaining capacity of metakaolin, which prevents rapid drying of concrete surface, as well as reducing cement consumption as most influential component on this parameter by replacing it with metakaolin. In addition, as shown in [61, 174], introduction of reactive mineral admixtures leads to a decrease in concrete shrinkage due to the absorption binding of water (as well as due to organizing the orderly structure of cement paste and reducing technological damage). Also, partial replacing of Portland cement with metakaolin leads to the reduction of the cement paste open porosity (see subsection 4.3.3), which means that the intensity of moisture exchange between concrete and environment decreases.

Crack resistance Parameters

Based on concrete strength and deformation characteristics given in Table 4.37, crack resistance of concrete can be estimated in accordance with known equations. The ratio $(f_t/f_c)\cdot 100\%$ is one of the criteria for concrete crack resistance [106, 175]. This criterion increase indicates growth of concrete crack resistance. Calculated values, listed in Table 4.38, are obtained on the basis of experimental data: the values are almost equal for different concrete compositions and vary at 4–6% on average, which is typical for high-strength concrete [171].

As noted above, concrete crack resistance is largely characterized by its ultimate tensile deformation, or ultimate elongation ε_{lim}. The following formulas were proposed for its evaluation:

$$\varepsilon_{lim} = f_t/E, \qquad (4.68)$$

$$\varepsilon_{lim} = f_b/E \qquad (4.69)$$

The calculated values of ε_{lim} are given in Table 4.38.

Table 4.38. Calculated Criteria of Concrete Crack Resistance.

No.	τ, days	$f_t/f_c \cdot 100\%$	$\varepsilon_1 = f_t/E \cdot 10^5$, MPa	$\varepsilon_2 = f_b/E \cdot 10^5$, MPa	$\varepsilon_3 = f_{st}/E_d \cdot 10^5$, MPa	$T = f_{st}/\varepsilon_s \cdot 10^{-5}$, MPa
	3	6.85	-	-	-	-
1	7	6.07	7.93	12.43	7.14	-
	28	5.35	7.61	12.62	8.04	7.57
	3	7.30	-	-	-	-
2	7	6.51	7.02	11.11	7.06	-
	28	5.52	7.30	12.07	8.21	10.54
	3	6.36	-	-	-	-
3	7	5.94	6.86	12.60	7.38	-
	28	4.93	7.04	13.86	8.74	17.99
	3	5.68	-	-	-	-
4	7	4.71	7.62	14.84	8.81	-
	28	4.47	8.05	14.28	9.99	30.01

Concrete tensile strength coincides quite closely with the ratio of splitting tensile strength to the dynamic modulus of elasticity, E_d [15]. The values of splitting tensile strength can be estimated as [171]:

$$f_{st} = 0.055(10f_c)^{2/3} \qquad (4.70)$$

and the value of the dynamic modulus of elasticity—by the formula [177]:

$$E_d = 4f_c/1(1 + 0.07 f_c) \qquad (4.71)$$

Values calculated by Eqs. (4.70) and (4.71) are given in Table 4.37, the value of conditional elongation—in Table 4.38. As shown in Table 4.38, the values of ultimate elongation increase with increasing the concrete age and at introduction of metakaolin. The values of ε_1 and ε_2 are similar. According to experimental data, the value of ε_{lim} in concrete is $(5 \ldots 15) \cdot 10^{-5}$ [171]. For the design of structures, the value of ε_{lim} is usually taken as $1 \cdot 10^{-4}$.

Along with direct measurement of separate properties, it is also possible to estimate general parameters to prognose concrete crack resistance. A modulus of crack formation, caused by shrinkage deformations, ε_s, was proposed [177]:

$$T = R_t/\varepsilon_s \qquad (4.72)$$

It is calculated for all the batches (Table 4.37).

Reduction of shrinkage deformations at partial replacement of cement with metakaolin and reduction of W/B leads to increase in T.

Thus, adding metakaolin leads to improvement of concrete strain properties, but at the same time the ratios between tensile strength and compressive strength, as well as bending (flexural) to compressive strength, decrease. The values of the calculated crack resistance criteria are given in Table 4.38, and indicate an increase in crack resistance with increasing content of metakaolin.

4.5.4 Durability Properties of Concrete

The required concrete durability is provided comprehensively by the design of optimal structure and composition, construction technology, protective measures [178, 179]. The important properties of concrete, which determine its durability, is frost resistance and watertightness.

Frost Resistance

It is known that concrete frost resistance is determined mainly by concrete pore structure nature: volume and size of pores, as well as the ratio of conditionally closed pores formed by contraction and air entrainment and open pores saturated with freezing water [180]. Self-leveling concrete in comparison to low-slump concrete has higher capillary pores content due to its significant water content. Introduction of superplasticizer, decreasing the water content, leads to a natural decrease in capillary porosity, compaction of cement paste and concrete structure and a corresponding increase in concrete frost resistance [13, 86]. However, there are the data on the reduction of concrete frost resistance due to deterioration of its pore structure at increase in pores number > 500 μm [86]. There is contradictory data on the effect of "superplasticizer-active mineral admixture" complexes on frost resistance. However, most researchers are inclined to prove that introduction of such additives in cement systems improves concrete frost resistance by reducing the pore size [94], but this indicator is very sensitive to additive overdoses [86, 87, 181].

Results of determining porosity of cement paste with a complex additive based on metakaolin shown in subsection 4.3.3 indicate increase in homogeneity of pore sizes and decrease in their average diameter at introduction of metakaolin. It gives us a reason to expect an improvement in the frost resistance of concrete with metakaolin admixture.

Compositions of concrete specimens, for which frost resistance was determined, are given in Table 4.39.

Table 4.39. Concrete Compositions.

No.	W/C	W/B	Basic components, kg/m³				Additives, % Binder / kg/m³	
			Water	Cement	Sand	Crushed stone	SP	MK
1	0.40	0.40	200	500	610	1105	0.50/2.50	–
2	0.42	0.40	200	475	608	1101	0.55/2.75	5/25
3	0.44	0.40	200	450	606	1097	0.75/3.75	10/50
4	0.38	0.34	170	450	634	1149	1.35/6.75	10/50

Frost resistance was obtained for $10 \times 10 \times 10$ cm samples by the second accelerated method [182, 183]. Frost resistance coefficient was applied as a criterion. The criterion is a ratio of the compressive strength of the main samples after a certain number of freezing and thawing cycles (f_o) to the control samples strength (f_c). The required number of freezing and thawing cycles for concrete composition No. 1 was taken according to [182, 183] and in accordance with available data [86]. For compositions 2–4 hypothesis of increasing frost resistance at introduction of metakaolin was considered. Figure 4.56 presents graphical dependences of K_F on the number of freezing—thawing cycles.

The experimental data is well approximated by the following equation:

$$K_F = 100 \cdot \sqrt{1 - X^2 / A^2} \qquad (4.73)$$

where X is a number of alternate freezing-thawing cycles;

A is a coefficient, corresponding to ultimate number of freezing-thawing cycles after which destruction occurs.

Values of coefficient A for concrete specimens, tested according to the second method, are given in Table 4.40. Taking $K_F = 95\%$, the ultimate number of freezing–thawing cycles and a corresponding frost resistance class were determined.

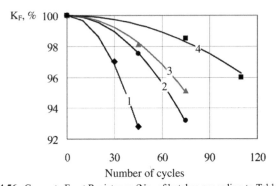

Figure 4.56. Concrete Frost Resistance (No. of batches according to Table 4.38).

Table 4.40. Experimental Determination of Frost Resistance of Concrete.

No.	f_c, MPa	A	Number of cycles at $K_F = 95\%$(2nd method)	Frost resistance class
1	49.6	120	37	150
2	54.5	205	64	200
3	60.7	240	75	300
4	78.7	405	126	400

As evidenced by the data of Fig. 4.58 and Table 4.40, an increase in the metakaolin portion in binder leads to an improvement of frost resistance; it confirms the hypotheses, given above. Introduction of metakaolin provides a reduction in pore size and, accordingly, a low permeability of cement paste and concrete. It is known that at W/C (W/B) decreasing and hydration degree increasing, the pore size decreases and less water will freeze at a constant temperature [116]. It is consistent with the obtained experimental data (concrete compositions 3 and 4, Table 4.40).

Traditional experimental methods for determining frost resistance do not enable us to predict frost resistance at the concrete proportioning stage. Therefore, a number of calculation methods for assessing frost resistance based on the factors that mostly affect it were developed.

It is known that capillary porosity has the main effect on frost resistance. Therefore, the attempt to link frost resistance with capillary porosity value was made by G. Gorchakov [184]. For normal hardening of concrete prepared on standard materials, the following dependence was suggested:

$$F = K \, (P_i - P_c) \, n \qquad (4.74)$$

where F is the number of freezing thawing cycles (which causes a certain degree of destruction);

K, n, P_i are parameters that depend on the materials quality, concrete composition, production factors;

P_c is capillary porosity, %:

$$P_c = (W-0.42\alpha C)/1000 \tag{4.75}$$

Statistical processing allowed to specify Eq. (4.74) into an empirical one:

$$F = (14-P_c)^{2.7} \tag{4.76}$$

and propose a method for determining concrete composition with the required frost resistance [184]. This method does not take into account the effect of the main structural parameter—the ratio of closed and open capillary pores, on concrete frost resistance, which limits application of the proposed dependence (4.74) by concrete without entrained air. Knowing values of α, W, C, the value of P_c was obtained by Eq. (4.75) and the F-value - by Eq. (4.76). The results are given in Table 4.41.

Frost resistance class F, can be predicted at a given W/C or C/W at a specified value of freezing-thawing cycles by the same formula that was used for calculation of concrete compressive strength f_c [185]:

$$F = Af_c(C/W-0.5) \tag{4.77}$$

where f_c is the cement grade;

A is coefficient, its recommended value for the control composition is 0.45, and for other compostions it is recommended to specify the value using a special nomogram, which takes into account the influence of W/C, water demand, type of admixtures, etc.

Introduction of active mineral admixtures leads to increase in concrete water consumption, capillary porosity growth and, subsequent decrease in frost resistance [185]. However, an increase in frost resistance is observed at adding superplasticizers. This phenomenon is explainded by redistribution of porosity in the direction of gel pores growth, as shown in subsection 4.3.3. Substituting W/B instead of W/C, the F-value is calculated by Eq. (4.76) for each batch (Table 4.41).

The following formula was proposed for determining frost resistance, based on statistical processing of a large number of experimental data [178]:

$$F = A_1 \cdot f_c^{A_2} \cdot e^{A_3 V_a} \tag{4.78}$$

where is concrete compressive strength at 28 days, MPa;

A_1, A_2, A_3 are constants;

V_a is the amount of entrained air, %.

For fresh concrete with high workability $A_1 = 0.34$; $A_2 = 1.68$; $A_3 = 0.35$. Frost resistance for the studied concrete according to the above equations is presented in Table 4.41.

It should be noted that the F-values determined by Eq. (4.76) are inflated, especially at low W/C (batch 4). Concrete frost resistance grade determined by Eq. (4.75) is also slightly inflated compared to the experimental data. Among all the above equations, the most accurate prediction of frost resistance can be done

Table 4.41. Determination of Frost Resistance of Concrete by Calculation Method.

No.	Frost resistance class	Initial data for calculation							
		Eq. (4.76)				Eq. (4.77)		Eq. (4.78)	
		A	P_c, %	F, cycles	Grade	F, cycles	Grade	F, cycles	Grade
1	150	0.767	3.893	516	500	424	400	240	200
2	200	0.724	4.796	401	400	408	400	281	200
3	300	0.732	4.628	421	400	426	400	337	300
4	400	0.767	0.893	1041	1000	534	500	521	500

using Eq. (4.77), although at low values of W/C (W/B) coefficients A_1 and A_2 should be adjusted.

Thus, the studied concrete can be classified as raised (F150–F300) and high (F300–F500) frost resistance [178].

Watertightness of Concrete

Concrete watertightness also significantly depends on its pore structure characteristics, in particular the volume and size of the through pores, which are actually ways to water filtration. Sedimentation processes of fresh concrete, attributable to self-leveling concrete, are important in the pores formation [90, 151]. Introduction of superplasticizer and metakaolin allows retarding the sedimentation processes; it can be assumed that it increases concrete watertightness. The most active part of pore space characterizes the volume of open pores available for water absorption. Therefore, according to theoretical assumptions, a clear correlation between open porosity and concrete watertightness should be expected.

In order to confirm this hypothesis, watertightness, and water absorption of concrete specimens' were determined. Porosity parameters were calculated according to the method described in [116]. Water absorption was obtained using cubes with edge size of 10 cm, watertightness at 28 days—using cylinders with a diameter and height of 15 cm. The values of these parameters are given in Table 4.42 and Fig. 4.57.

As can be seen from the above data, there is a direct correlation between concrete water absorption and watertightness. Increase in metakaolin content in binder leads to decrease in average diameter of pores, improving their homogeneity, reducing water absorption and subsequent increasing concrete water tightness (batches 1–3, Table 4.42). Decrease in W/B, which is actually a decrease in water content at other stable conditions, leads to decrease in water absorption and increase in watertightness (points 3 and 4, Table 4.42).

Forecasting concrete watertightness at proportioning stage also requires empirical equations; one of the them was represented by V. Sizov [186]:

$$W = Af_c(C/W - 0.5) \tag{4.79}$$

Table 4.42. Water Absorption and Watertightness.

No.	Porosity parameters		Water absorption, % experimental values		Watertightness				
					experimental values		calculated values		
	α	λ	by weight	by volume	Pressure, MPa	Class	A	Pressure, MPa	Class
1	0.17	1.98	4.16	5.89	0.84	8	0.840	0.84	8
2	0.29	1.68	2.17	3.11	1.15	10	1.030	1.08	10
3	0.32	1.84	1.26	2.90	1.38	12	1.200	1.40	14
4	0.45	1.57	0.85	1.76	1.55	14	1.290	1.51	14

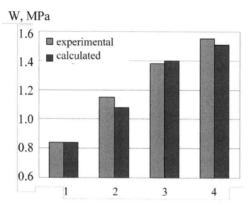

Figure 4.57. Watertightness of Concrete Specimens (according to Table 4.42).

where coefficient A is taken equal to 1 for the control batch, followed by adjustment by a special nomogram, as in Eq. (4.79). As in case of determining the frost resistance, it is believed that the effect of active mineral admixtures is negative, as a result, the calculated values of concrete watertightness with the addition of metakaolin were underestimated. Therefore, in order to eliminate this discrepancy, the value of the adjustment factor A_{Dmin} was included. Its value should be clarified as the number of statistics increases. The experimental and calculated values of watertightness are presented in Table 4.42. The conducted studies indicate an increase in concrete frost resistance and water resistance with the introduction of metakaolin.

4.6 Other Types of Concrete Admixtures Containing Metakaolin

Application of other Types of Admixtures for Concrete containing Metakaolin

As shown in subsection 4.4.1, the requirements for workability retention are put forward to self-leveling concrete. For concrete batches of the above compositions, the retention time value averages about two hours and naturally decreases with ambient temperature growth.

The compressive strength of studied concrete at 28 days in average was 50–80 MPa. However, according to advanced ideas, high-strength concrete has compressive strength of more than 60 MPa [91, 187, etc.]. High-strength concrete is defined as concrete with a specified compressive strength of 55 MPa or higher [188], whereas at many markets today, concrete with the compressive strength more than 69 MPa (10,000 psi) is routinely produced [188]. Therefore, further studies were aimed at increasing the workability retention of fresh concrete and the compressive strength of hardened concrete. There is available data on the effectiveness of the combined use of metakaolin and gypsum dihydrate and hemihydrate from the viewpoint of ettringite formation at early hardening stages, which allows for an increase in strength and significantly reduces shrinkage deformations [189–191]. It is possible to achieve this goal by additional fresh concrete plasticization and adding retarding agents, as well as by reducing concrete water demand. Reduction of water demand is possible due to partial replacement of metakaolin with another material with lower specific surface area, like gypsum (dihydrate).

Lignosulfonate technical (LST) was added to the composite metakaolin-based modifying admixture for additional mixture plasticization and retarding effect. According to known data, the addition of LST at a dosage higher than 0.5% by cement weight leads to retardation in the concrete structure-forming processes and overall decreasing strength and strain properties [192], so a suggested dosage less than 0.25% by cement weight is common. At the same time, at combined use of LST and NF type SP workability of fresh concrete either remains the same or slightly increases [111, 192]. In order to avoid the negative effects of LST, its dosage usually does not exceed 0.1% by cement weight.

The water-binder ratio was fixed from the point of providing fresh concrete workability at the initial time at the level of 22 ± 2.5 cm (S15).

Control concrete batch contained NF type SP and SP NF + LST. The following initial materials were used:

- Portland cement of CEM I42,5;
- quartz sand with fineness modulus $M_f = 1.80$ and 2.40;
- granite crushed stone, mixture of fractions: 5–10 mm (30%) and 10–20 mm (70%);
- composite admixture consisting of NF type superplasticizer (SP); metakaolin (MK), $s = 1670$ m²/kg; calcium sulphate dihydrate (G), $s = 300$ m²/kg; LST = 0.1% by cement weight.

The concrete compositions and research results are given in Table 4.43 and in Figs. 4.58–4.60.

During the experiment, fresh concrete workability was determined by the standard cone slump and flow spread diameter D_f immediately after mixing, in 1 h and after 2 h. According to the obtained results, workability retention τ^r_{22-18} was determined (see subsection 4.4.1) as the time during which the fresh concrete

Table 4.43. Compositions of Concrete and Research Results.

No.	1	2	3	4	5	6	7	8	9	10	11	12	13
W/C	0.38	0.38	0.40	0.37	0.37	0.36	0.38	0.37	0.36	0.34	0.40	0.38	0.34
W/B	0.38	0.38	0.36	0.34	0.34	0.33	0.35	0.34	0.33	0.31	0.36	0.35	0.31
C, kg/m³	530	530	500	500	500	500	500	500	500	500	500	500	500
Composite admixture %C / kg/m³	-	-	10/50	10/50	10/50	10/50	10/50	10/50	10/50	12/60	10/50	10/50	10/50
SP, %C	0.6	0.6	1.2	1.2	1.4	1.4	1.2	1.2	1.2	1.4	1.2	1.2	1.4
MK, %C	-	-	8.8	8.8	8.6	8.6	8.3	8.3	6.8	8.2	6.8	6.8	6.6
G, %C	-	-	-	-	-	-	0.5	0.5	2.0	2.4	2.0	2.0	2.0
LST, %C	-	0.1	-	0.1	-	0.1	0.5	0.1	0.1	0.1	-	0.1	0.1
t_{22-18}, h — 0	2.45	2.45	2.25	5.20	1.85	2.75	2.75	4.55	3.35	0.50	2.8	3.55	0.50
SI/D₀, cm, after τ (h after mixing) — 0	22.5/49	22.5/53	22/42	23.5/50	22.5/42	23/48	22.5/46	23.5/49	24.5/54	20/37	25/53	25.5/55	22/45
SI/D₀ — 1	22/42	22/52	21.5/39	23.5/43	21/36	22/46	21.5/43	23.5/47	24/54	18/20	24.5/52	25/55	10/25
SI/D₀ — 2	20/38	20/44	19/33	22.5/34	18/31	20.5/35	20/38	23/45	23/53	-	23/52	24/54	-
tg φ (4.4.2), h after mixing — 0	0.31	0.28	0.38	0.26	0.36	0.29	0.30	0.27	0.20	0.54	0.19	0.16	0.36
tg φ — 1	0.38	0.31	0.44	0.30	0.50	0.35	0.40	0.28	0.22	1.20	0.21	0.18	1.60
tg φ — 2	0.42	0.37	0.67	0.44	0.77	0.55	0.53	0.31	0.26	-	0.27	0.22	-
f_c, MPa, at the age, days — 3	38.7	38.5	49.1	48.9	40.5	52.5	48.9	45.8	49.0	48.3	29.9	33.2	42.2
f_c — 7	56.3	52.2	66.7	66.5	67.3	70.2	66.5	63.3	63.9	72.5	47.8	48.5	62.5
f_c — 28	69.5	66.7	85.8	85.9	86.1	87.8	83.3	77.0	86.9	92.7	62.3	68.4	86.5
f_c — 90	-	-	-	-	-	-	-	-	93.1	99.5	-	-	94.9

Note. Quartz sand with fineness modulus of $M_f=1.8$ was applied for batches 12 and 13.

slump will decrease by 4 cm from the initial (i.e., the average value of nearest lower workability class). The fresh concrete internal friction angle (segregation parameter tg φ) was also determined (see subsection 4.4.2). Hardened concrete compressive strength was determined at 3, 7 and 28 days (for certain batches at 90 days).

As it can be seen from Fig. 4.58a, adding LST in concrete with a composite metakaolin-based additive modifier, increases fresh concrete workability retention (batches 4, 8, 12). Whereas at adding LST to concrete batch without metakaolin admixture does not affect the retention (batches 1 and 2). Although it should be noted that the flow spread of fresh concrete with SP and LST is higher even at the equal slump (batches 1 and 2, Table 4.43). The higher the gypsum content in composite admixture is, the lower the LST effect on workability retention is.

As superplasticizer content grows, the workability retention decreases. The higher the gypsum content is, it occurs more intensively. It is due to the achievement of equal workability and the intensification of hydration processes at the corresponding reduction in fresh concrete water content (Fig. 4.58b). Increase in gypsum content in composite admixture leads to a slight growth of water retention for fresh concrete without LST (Fig. 4.58, batches 3,7,11). Whereas at adding 0.1% of LST the dependence is the opposite: the higher the gypsum content, the lower is the water retention value. Increasing the total amount of modifier from 10 to 12% by cement weight leads to a water retention reduction. It can be explained by corresponding the decrease in W/C and the lack of free water for preserving the workabilty (Fig. 4.58c,d).

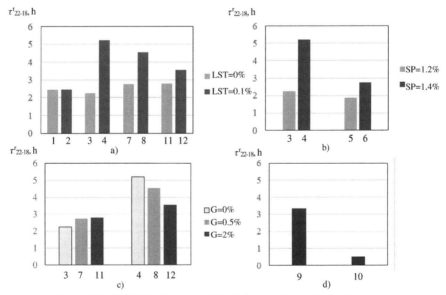

Figure 4.58. Workability Retention of Fresh Concrete Depending on Dosage of:
a) LST admixture; b) SP; c) dihydrate calcium sulphate (G); d) composite admixture (SP + MK+G). Designation according to Table 4.43.

The internal friction angle (tg φ) for all types of concrete increases with time. For fresh concrete containing LST, this process retards in time comparing to mixtures without this admixture (Table 4.43, batches 1 ... 4, 7 and 8). The SP content growth leads to an intensive increase in the value of tg φ due to reduction of water layers between the solid particles (batches 5 and 6). Increase in the gypsum content in composite admixture leads to retardation of tg φ growth (batches 4, 8 and 12), and with increasing the total content of metakaolin modifier a reverse process is observed (batches 9 and 10).

Comparing the compressive strength of concrete specimens with and without LST at the same age, almost equal values or even exceeding by 5–10% are evident for specimens with LST. It can be explained by a lower W/C due to the additional plasticization of fresh concrete (Fig. 4.59a). Increasing SP in the composite admixture causes subsequent compressive strength growth due to a reduction of water demand and W/C by 10–15% (Fig. 4.59 b). The higher the dosage of metakaolin replacement with gypsum is the lower is its compressive strength (Fig. 4.59c). The overall increase of composite admixture content from 10 to 12% (batches 9 and 10, Table 4.43) are due to W/C reduction from 0.36 to 0.34 which leads to compressive strength growth (Fig. 4.59 d).

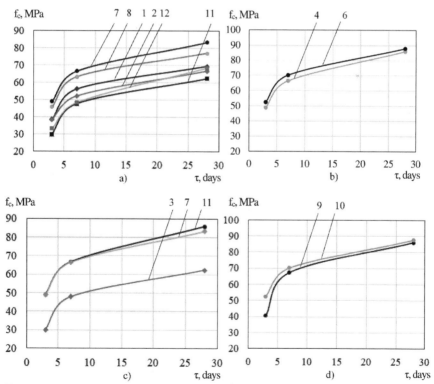

Figure 4.59. Kinetics of Concrete Compressive Strength Depending on: a) presence of LST admixture; b) dosage of SP; c) dosage of dihydrate calcium sulphate (gypsum); d) total dosage composite additive (SP+MK+G). Designation according to Table 4.43.

Thus, the additional modification of concrete with LST admixture enables us to increase the fresh concrete workability, and introduction of calcium sulphate dihydrate leads to an increase in concrete strength. Reduction of the mixture water demand due to either increasing the superplasticizer dosage or replacing part of the metakaolin with gypsum leads to the hydration processes intensification and the corresponding decrease of fresh concrete workability retention values.

4.7 Peculiarities of Industrial Manufacturing of Concrete Containing Metakaolin and Superplasticizer

4.7.1 Concrete Proportioning Method and Proportion Optimization

Self-leveling high-strength concrete mix proportion differs from that for normal strength concrete as chemical and mineral admixtures content and influence should be considered [167, 193].

Self- leveling High-strength Concrete Proportioning

For each batch (subsections 4.4.1–4.4.3) the compositions were calculated per 1m³. At the concrete proportioning stage the absolute volumes rule [88, 194] was applied.
The water demand is calculated as follows:

$$W = B \cdot (W/B) = X_1 \cdot X_2 \tag{4.79}$$

where B is binder (cementitious material) content, kg/m³(X_2, in Table 4.23);
W/B is the water-binder ratio (X_1, in Table 4.23).
The cement and metakaolin contents are obtained as:

$$C = B (1-X_3/100) \tag{4.80}$$

$$MK = B \cdot X_3/100 \tag{4.81}$$

The binder volume is found as

$$V_B = W + C/\rho_c + MK/\rho_{MK} \tag{4.82}$$

where ρ_c and ρ_{MK} are densities of cement and metakaolin, kg/m³ respectively.
According to the absolute volumes rule, the total aggregates volume is

$$V_A = 1000 - V_B \tag{4.83}$$

The sand volume portion in total volume of aggregates *r* for fresh concrete of high workability is determined using a nomogram, depending on cement (binder) content and sand fineness modulus [95]. Thus, at a binder content of 450 kg/m³ the sand volume share is 0.37, at 500 kg/m³ – 0.355 and at 550 kg/m³ – 0.34, respectively.

Sand (S) and crushed stone (CS) contents are determined as:

$$S = V_B \cdot r \cdot \rho_s \tag{4.84}$$

$$CS = V_B(1 - r)\rho_{cs} \tag{4.85}$$

where ρ_s and ρ_{cs} are densities of sand and crushed stone, kg/m^3.

Optimization of Self-leveling High strength Concrete Proportioning

For concrete proportion optimization was applied a method, which includes both deterministic and stochastic equations. Materials, described in subsections 4.4.1–4.4.3 were applied. A three-factor three-level plan, similar to a D-optimal one was used in the experimental work (Table 4.44). Quartz sand with fineness modulus of 1.5 was used.

Table 4.44. Factorial Plan of the Experiments.

Factors designation			Variation levels		
Designation	Name		Bottom (-1)	Mean (0)	Upper (+1)
X_1	Water-binder ratio (W/B)		0.34	0.37	0.40
X_2	Binder content (B), kg/m^3		450	500	550
X_3	Metakaolin portion (MK),% by binder content		5	10	15

Two sets of experiments were conducted: first one containing NF type SP (SP$_2$, see subsection 4.4.2) and second one with PC type SP. The SP content in each batch was selected to provide an equal slump of 22 ± 1.5cm (slump flow diameter varied from 400 to 550 mm). Batches compositions for the 1st and 2nd sets of experiments are shown in Table 4.45. To meet the self-leveling and self-compacting concrete design requirements [132], the coarse aggregate was limited to 1250 kg/m^3 (50% by solid volume) and 700 kg/m^3 (40% by mortar volume), see subsection 4.4.3.

At both sets of experiments the compressive strength and efficiency factor have been obtained according to Eq. (4.57). Additional tests for concrete of control compositions have been carried out to calculate the efficiency factor (see subsection 4.5.2, Table 4.33).

The experimental data were used to calculate the efficiency factor of metakaolin in SP$_1$ (NF type SP based) concrete (Table 4.34) and SP$_2$ (PC type SP based) concrete (Table 4.35) as it is shown in subsection 4.5.2. Regression equations coefficients and corresponding statistical (calculated) characteristics were obtained (see for PC type SP). Statistical models of SP content (in % by cementitious material mass), compressive strength at 28 days f_c and metakaolin efficiency factor F$_e$ calculated by Eq. (4.57) were obtained as follows (agglomerated data according to subsections 4.4.1, 4.5.1 and 4.5.2):

- for NF type SP based concrete

$$SP_{NF} = 1.19 - 0.32x_1 - 0.26x_2 + 0.35x_3 - 0.09x_2{}^2 + 0.08x_3{}^2 - 0.11x_1x_3$$
$$- 0.07x_2x_3 \qquad (4.86)$$

$$f_{c1} = 71.3 - 9.3x_1 + 3.4x_2 + 4.3x_3 - 1.9x_2{}^2 - 1.7x_3{}^2 + 1.2x_1x_3 + 1.5x_2x_3 \quad (4.87)$$

$$F_{e1} = 1.43 - 0.22x_1 + 0.05x_2 + 0.17x_3 - 0.06x_2{}^2 - 0.02x_1x_2 + 0.04x_2x_3 \quad (4.88)$$

- for SP of PC type based concrete:

$$SP_{PC} = 0.26 - 0.06x_1 - 0.06x_2 + 0.05x_3 + 0.04x_1{}^2 + 0.03x_2{}^2 \qquad (4.89)$$

$$f_{c2} = 77.0 - 9.4x_1 + 2.6x_2 + 3.9x_3 - 1.4x_2{}^2 - 1.8x_3{}^2 + 0.8x_1x_3 + 1.6x_2x_3 \quad (4.90)$$

$$F_{e2} = 1.48 - 0.21x_1 - 0.05x_2 + 0.15x_3 + 0.03x_1{}^2 - 0.03x_3{}^2 + 0.03x_1x_2 \quad (4.91)$$

Following the experimental data and statistical processing, the obtained regression equations and diagrams (see subsections 4.4.1, 4.5.1 and 4.5.2) can be used for cost-effective concrete design and solving concrete composition optimization problems.

Minimal cost is considered to be one of the most common optimization criteria in concrete design. It saves energy and the most expensive components for concrete production like cement, mineral and chemical admixtures. Therefore, concrete composition optimization is directly related to minimization of cement, superplasticizer and metakaolin contents. As the aggregates price is relatively low, their contribution to the overall concrete cost is negligible for all concrete compositions. In the frame of this study concrete manufacturing costs were assumed to be equal for all concrete compositions. For concrete cost minimization the following objective function C_B is used:

$$C_B = C_C \cdot C + C_{SP} \cdot SP + C_{MK} \cdot MK \rightarrow min \qquad (4.92)$$

where C_C, C_{SP}, C_{MK} are costs of cement, superplasticizer and metakaolin respectively, USD/kg;

C, SP, MK – content of cement, superplasticizer and metakaolin respectively, kg/m^3.

Using the regression equations (4.86) – (4.91) and the objective function (Eq. 4.92), dependence of binder cost on metakaolin content, minimal metakaolin and superplasticizer contents for specified concrete manufacturing conditions are obtained. Concrete compressive strength and metakaolin efficiency factor are the optimization constraints.

The method is explained by giving an example of concrete composition design under the following constrains: minimum concrete compressive strength at 28 days is 80 MPa ($f_c \geq 80$ MPa); efficiency factor $F_e > 1.2$; the concrete cost is minimum.

A graphic approach can be used to find the regression equations' solutions. Following the diagrams given in subsection 4.5.1, compressive strength of

80 MPa is achieved at 28 days if W/B = 0.34–0.35 for SP_{NF} or if W/B = 0.34–0.37 for SP_{PC}.

Initially the x_1 (W/B) value was stabilized at the bottom level -1 (corresponding to the natural value of 0.34) and subsequently it was moved with spacing of 0.33 (natural value of 0.01). For each x_1 value a constraint function was determined at f_c=80 MPa, derived from Eqs. (4.87) and (4.90). Thus, at $x_1 = -1$ ($X_1 = 0.34$) the equations have the following form:

for SP_{NF} (Eq. 4.87):

$$80 = 71.3 - 9.3(-1) + 3.4x_2 + 4.3x_3 - 1.9x_2^2 - 1.7x_3^2 + 1.2(-1)x_3 + 1.5x_2x_3,$$

$$0.6 + 3.4x_2 + 3.1x_3 - 1.9x_2^2 - 1.7x_3^2 + 1.5x_2x_3 = 0$$

and for SP_{PC} (Eq. 4.90):

$$80 = 77.0 - 9.4(-1) + 2.6x_2 + 3.9x_3 - 1.4x_2^2 - 1.8x_3^2 + 0.8(-1)x_3 + 1.6x_2x_3,$$

$$6.4 + 2.6x_2 + 3.1x_3 - 1.4x_2^2 - 1.8x_3^2 + 1.6x_2x_3 = 0$$

The graphical approach for searching the feasible solutions of the problem, obtained by the f_c= 80 MPa level curve imposition and efficiency factor level curves, is presented in Fig. 4.60. According to Fig. 4.60, in the cross points of the level curve f_c= 80 MPa with the experimental area limits and the efficiency factor curves the values of factors x_2 (binder content) and x_3 (metakaolin portion) are determined. Due to non-linear programming approach, if the level curve f_c= 80 MPa is parabolic (Fig. 4.60a), its inflection point is considered to be the minimal cost point [139, 141]. If the level curve f_c= 80 MPa can be approximated by a linear function (Fig. 4.60b), the minimal cost is achieved at the end point of the line segment [195].

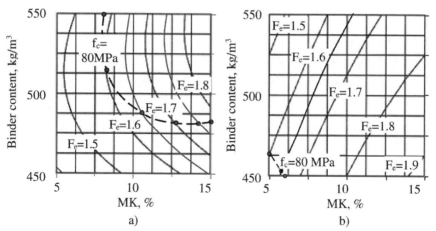

Figure 4.60. Graphical Approach for Searching Feasible Solutions of: a) SP_1 (NF) based concrete, W/B = 0.34; b) SP_2 (PC) based concrete, W/B = 0.34.

Table 4.45. Design Data for Searching the Optimal Values of the Investigated Factors.

Parameters		SP$_1$ NF (naphthalene formaldehyde)						SP$_2$ PC (polycarboxylate)		
Efficiency factor value		1.58	1.60	1.65	1.70	1.75	1.77	1.60	1.65	1.66
W/B	symbolic x_1	−1	−1	−1	−1	−1	−1	−1	−1	−1
	natural X$_1$	0.34	0.34	0.34	0.34	0.34	0.34	0.34	0.34	0.34
Binder, kg/m³	symbolic x_2	1	0.32	−0.25	−0.38	−0.38	−0.37	−0.74	−0.95	−1
	natural X$_2$	550	516	488	481	481	482	463	453	450
MK, %	symbolic x_3	−0.40	−0.36	0.13	0.57	0.87	1.00	−1.00	−0.85	−0.80
	natural X$_3$	8.0	8.2	10.7	12.9	14.4	15.0	5.0	5.8	6.0
SP	portion, % by binder content	1 02	1.27	1.63	1.89	2.07	2.15	0.29	0.32	0.33
	content, kg/m³	5.6	6.6	8.0	9.1	10.0	10,4	1.35	1.46	1.49
	cost, USD/kg	1.59	1.59	1.59	1.59	1.59	1.59	16.15	16.15	16.15
C	content, kg/m³	506	474	436	419	412	409	440	426	423
	cost, USD/kg	0.11	0.11	0.11	0.11	0.11	0.11	0.11	0.11	0.11
MK	content, kg/m³	44	42	52	62	69	72	23	26	27
	cost, USD/kg	0.48	0.48	0.48	0.48	0.48	0.48	0.48	0.48	0.48
Total cost of cement and admixtures, USD/m³		85.67	82.86	83.21	85.48	90.26	94.29	81.24	82.90	83.47

Converting symbolic values of factors into natural values by applying the following formula the corresponding W/B, binder and metakaolin contents can be determined:

$$X_i = \Delta X_i \cdot x_i + X_{i0} \tag{4.93}$$

Here x_i is the symbolic value of the factor;

ΔX_i is factor variation interval;

X_{i0} is mean variation level (zero point) of the factor.

By substituting the factors values in Eqs. (4.86) and (4.89), the superplasticizer content is obtained. Then the binder cost is found according to Eq. (4.92). An example of design data and calculation results are shown in Table 4.45.

Based on the data given in Table 4.45, dependencies of binder cost on metakaolin efficiency factor are obtained (Fig. 4.61). As it follows from the figure, the cross points can be approximated either by parabola (Fig. 4.61a) or linear function (Fig. 4.61b) at squared approximation reliability R higher than 0.95. In the first case the parabola peak coordinates correspond to the minimum binder cost.

For NF type SP-based concrete considering that derivative $\dfrac{\partial y}{\partial x}\left(\dfrac{\partial C_B}{\partial F_e}\right)$ in this point is equal to zero, the parabola peak coordinates are calculated as follows:

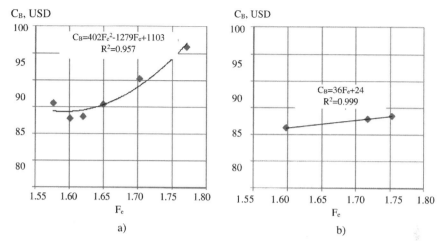

Figure 4.61. Dependence of binder cost on metakaolin efficiency factor: a) SP$_1$ based concrete, W/B = 0.34; b) SP$_2$ based concrete, W/B = 0.34.

$$\frac{\partial C_B}{\partial F_e} = 402 \times 2 \times F_e - 1279 = 0 \tag{4.94}$$

where F_e = 1.59; C_B = 84.09 USD.

For PC type SP–based concrete the cross points are approximated by a linear function and the minimal cost is in the end point, where F_e = 1.6; C_B = 81.24 USD. Efficiency factor values in both cases are similar and exceed 1.2, demonstrating that metakaolin is a high-efficiency mineral admixture. Minimum cost values for the rest of W/B ratios were calculated using the same method. The minimum cost values are approximated by linear functions (Fig. 4.62). The minimum binder cost is provided at the point where W/B is minimal. It is because at W/B growth the binder content should be increased, subsequently yielding higher metakaolin and superplasticizer contents for providing equal strength. It should also be mentioned that the binder cost is lower when PC type SP is used, because the required strength in this case is achieved at a lower binder content and metakaolin portion in it. For calculating the X_2 and X_3 values, a set of level curves equations was solved for R_c = 80 MPa, F_e = 1.59 for SP$_1$ based concrete and R_c = 80 MPa, F_e = 1.6 for SP$_2$ based concrete.

$$\begin{cases} 0.6 + 3.4x_2 + 3.1x_3 - 1.9x_2{}^2 - 1.7x_3{}^2 + 1.5x_2x_3 = 0 \\ 0.06 + 0.07x_2 + 0.17x_3 - 0.06x_2{}^2 + 0.04x_2x_3 = 0 \end{cases} \tag{4.95}$$

$$\begin{cases} 6.4 + 2.6x_2 + 3.1x_3 - 1.4x_2{}^2 - 1.8x_3{}^2 + 1.6x_2x_3 = 0 \\ 0.12 - 0.08x_2 + 0.15x_3 - 0.03x_3{}^2 = 0 \end{cases} \tag{4.96}$$

The values of x_2 and x_3 can be determined from the diagrams (Fig. 4.61).

Thereby for NF type SP based concrete: x_2 = 0.64 (Binder = 532 kg/m³), x_3 = –0.41 (MTK = 8% by binder weight) at x_1 = –1 (W/B = 0.34);

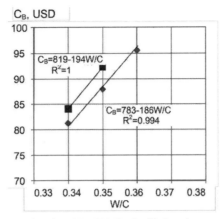

Figure 4.62. Minimum Cost Values for R28 = 80 MPa: 1 – SP_1 based concrete; 2 – SP_2 based concrete.

Table 4.46. Minimal Cost Compositions for C70 Concrete Class.

SP type	W/B	Concrete composition, kg/m³					
		W	C	S	CS	MK	SP
SP NF type	0.34	181	490	589	1156	42	6.1
SP PC type	0.34	157	440	669	1203	23	1.1

For SP PC type based concrete: $x_2 = -0.74$ (Binder = 463 kg/m³), $x_3 = -1$ (MTK = 5% of binder weight), at $x_1 = -1$ (W/B = 0.34).

The superplasticizer content is calculated from Eqs. (4.86) and (4.89). The concrete composition can be calculated according to the method described in subsection 4.7.1.a (Table 4.46).

The method proposed in this study can be also applied for other constraint functions. Solving the cost minimization problem is considered to be one of the ways to improve the sustainability of such materials as high strength concrete.

4.7.2 Industrial Manufacturing of Concrete Containing Metakaolin and Superplasticizer

Industrial Laboratory Tests

Laboratory tests have been carried out at the Kovalska Group (Kyiv) plant for reinforced concrete structures, which is currently the largest manufacturer of construction materials in Ukraine. Table 4.47 shows results of testing concrete with a composite admixture that consisted of NF type superplasticizer and metakaolin.

The following initial materials were used for the experiments: Portland cement CEM I 42.5 N, quartz sand with fineness modulus of 1.6–1.8; granite crushed stone of fractions of 5–10 mm (30–35%) and 10–20 mm (65–70%). Composite admixture Geokon-12 (NF type SP + defoaming agent – 12%, MK

Table 4.47. Results of Industrial Laboratory Tests of Concrete with Composite Admixture that Consisted of Superplasticizer and Metakaolin (data from S. Kovalska group plant).

Parameters	Batch No.					
	Series No.1			Series No.2		
	1	2	3	4	5	6
1	2	3	4	5	6	7
Cement, kg/m^3	500	500	500	470	450	470
Quartz sand, kg/m^3	650	650	650	630	650	630
Crushed stone 5-10 mm, kg/m^3	300	300	300	345	350	345
Crushed stone 10-20 mm, kg/m^3	850	850	850	804	810	804
Water, l/m^3	168	200	196	258	200	200
Geokon, % cement weight/kg/m^3	-	4/20	6/30	-	11/50	11/53
SP NF type, % by cement weight	0.6	-	-	-	-	-
W/C	0.336	0.40	0.392	0.55	0.44	0.43
Slump, cm						
in 5 min	16	17–18	17–18	23	23	23
in 15 min				18	23	22
in 60 min	-	-	-	16	21	19
in 120 min				14	18	17
Concrete compressive strength, MPa						
- steam curing	-	-	-	18.4	37.1	42.6
- normal hardening						
3 days	-	-	-	16.7	30.5	41.7
7 days	35.1.	40.4	50.3	27.4	43.6	57.8
28 days	47.5	-	76.0	34.3	54.6	65.7
Watertightness	W4	-	W10	-	-	-

with specific surface of 1.670 m^2/kg – 88%) was used for mixtures of series No.1, and Geokon G-14 (NF type SP + defoaming agent – 14% and MK – 86%) was used for batches of series No.2.

As can be seen from the experimental results of series No.1, introduction of the composite admixture even at increasing W/C allows us to increase compressive strength, which at 28 days can reach 160% (composition No.3) compared to control batch No.1, as well as an increase in watertightness (from W4 to W10). Compared to concrete without additives (batch 4), the introduction of the composite admixture allows us to achieve a significant increase in strength, while reducing W/C. At the same time, the fresh concrete workability retention increases, which is consistent with experimental data obtained in subsection 4.4.1.

Based on the results of experiments conducted at the industrial laboratory, the concrete batch with composite admixture containing superplasticizer and metakaolin was manufactured at the Kovalska group (Kyiv, Ukraine) plant.

The composition of the admixture is NF type SP + defoaming agent –14% (by weight), metakaolin - 86% (by weight). The compostion was proportioned for manufacturing reinforced concrete staircases cast in situ.

Results of Testing Industrial Batch of Self-compacting High-strength Concrete

Concrete batch (batch volume - 5 m³, designed concrete class C45) was manufactured and placed intoreinforced concrete staircase mould. The specimens were prepared at the laboratory of Kovalska Industrial-Construction Group (Kyiv, Ukraline), using components for ordinary concrete production.

Characteristic of Raw Materials:

- Portland cement CEM I 42.5R: specific surface – 3.000 m²/kg, normal consistency – 25.6%, setting time: initial – 2 h 15 min, final – 3 h 05 min, strength at 28 days: bending– 5.3 MPa, compressive – 50.3 MPa;
- Quartz sand: absolute density – 2.620 kg/m³, bulk density – 1.495 kg/m³, fineness modulus – 1.7, dusty and clayey particle content – 0.5%;
- Granite crushed stone: fractions 5–10 mm (30%) and 10–20 mm (70%), absolute density – 2.650 kg/m³, bulk density – 1.350 kg/m³, dusty and clayey particle content – 0.2 %;
- Composite additive consisted of: metakaolin (specific surface – 1.670 m²/kg, Al_2O_3– 40.3%, SiO_2– 54.6%) and NF type superplasticizer, modified by defoaming agent.

Concrete composition:

- Cement – 550 kg/m³;
- Sand – 540 kg/m³;
- Crushed stone: 5–10 mm – 478 kg/m³; 10–20 mm – 784 kg/m³;
- Composite admixture of 14% SP and 86% MK – 55 kg/m³;
- Water – 185 *l*/m³.

Fresh concrete density – 2430 kg/m³.

The batch was mixed in forced action mixer duration 25 sec with subsequent mixing in truck concrete mixer duration 5 min (velocity 14 rotations per min), 10 min pause. Initial slump was 23 cm (see Table 4.48). Cubic specimens (10 cm edge) were moulded and tested to obtain compressive strength after steam curing and normal hardening at 3, 7, 28 and 56 days (Table 4.48).

Frost Resistance and Watertightness Tests of High-strength Concrete

Samples were moulded from the experimental industrial batch to investigate frost resistance and the watertightness of concrete at the industrial laboratory. The test results are given below. Cubic specimens of normal-weight ready-mixed

Table 4.48. Results of Industrial Testing of Self-compacting High-strength Concrete Specimens.

No.	Parameter	Units	Value
1	Initial workability	cm	23
2	Workability in 15 min	cm	21
3	Workability in 1 h	cm	19
4	Workability in 2 h	cm	19
5	Compressive strength after steam curing	MPa	44.5
6	Compressive strength after 3 days	MPa	40.5
7	Compressive strength after 7 days	MPa	51.5
8	Compressive strength after 28 days	MPa	59.8
9	Compressive strength after 56 days	MPa	62.9

concrete, designed to meet following parameters: SL5C45F300W8 (Slump higher than 20 cm, concrete grade C45, frost resistance F300, watertightness W4) have been tested.

The following specimens were moulded:

• for studying frost resistance: 18 cubes, 15 cm edge;
• for studying watertightness: 6 cubes, 10 cm edge.

Results of Frost Resistance Testing

The number of cycles of specimens alternate freezing and thawing—75 cycles according to the accelerated method (air tested, $t = -18°C$, NaCl solution at $C = 5\%$). No losses of mass and strength were observed.No external signs of deterioration were observed.

According to standard requirements, concrete specimens meet requirements to concrete frost resistance grade F300.

Results of Watertightness Testing at 28 Days

The specimens meet requirements to watertightness grade W10 (filtration coefficient $K_f = 6.09 \cdot 10^{-11}$ cm/sec).

Expected Economic Effect of Manufacturing Self-compacting High-strength Concrete

Cost of concrete components when composite additive based on based on NF type superplasticizer and metakaolin, $/m^3$ have been compared to those with SP-silica fume additive. The economic effect of high-strength concrete Sl5C60 with composite admixture containing superplasticizer and metakaolin was calculated for comparison to concrete of the same class with composite admixture based on NF type SP and silica fume (SF). Raw materials: Portland cement CEM I N, quartz sand (fineness modulus 2.4), crushed stone: fractions 5–10 mm (30%) and 10–20 mm (70%).

Both of concrete composite admixtures contain 10% (by weight) of NF type superplasticizer and either silica fume or metakaolin. As can be seen from Table 4.49, in order to achieve the same fresh concrete workability at constant W/B it is necessary to increase the modifier containing silica fume consumption, compared to that containing metakaolin. At the same time, the expected economic benefit is around 8.5 $ per 1 m³. Other items of expenditure remain equal.

Table 4.49. Economic Benefit of Self-compacting High Strength Concrete.

Component	Composite additive SP+SF			Composite additive SP+MK		
	Dosage, kg/m³	Cost, $/t	Cost, $/m³	Dosage, kg/m³	Cost, $/t	Cost, $/m³
Water	185	0.5	0.09	185	0.50	0.09
Cement	425	84.8	36.04	450	84.80	38.16
Sand	618	7.8	4.82	629	7.80	4.91
Crushed stone	1,120	8.5	9.52	1.114	8.50	9.47
Composite cementitious admixture Geokon (14%SP+86%MK)	–	–	–	50.00	410.20	20.51
Composite admixture (10%SP+90%SF)	75	415.4	31.16	–	–	–
Total	–	–	81.63	–	–	73.14
Economic benefit	–	–	–	–	–	8.49

5

Use of Complex Metakaolin Fly Ash and Blast Furnace Granulated Slag in Reactive Powder Concrete (RPC)

The most effective type of fine-grained self-compacting concrete, which has a high homogeneity, strength and deformability was developed in France in the later part of the 20th century, and is Reactive Powder Concrete (RPC). The typical compressive strength for such concrete is in the range from 150 to 200 MPa, which is several times higher than that for conventional concrete. Under special conditions such concrete can achieve an increase in strength up to 800… 810 MPa. Following the available data, along with high strength RPC which also has a high crack resistance, characterized by the ratio of compressive to flexural strength. This indicator for RPC is in the range from 3.5 to 5, while for traditional high-strength concrete it is 8… 10. Such a high performance can be useful from the viewpoint of achieving the high impact strength required for fortification and other protective structures. Combination of ultra-high strength and high deformability in RPC is provided by dispersed reinforcement by short steel fibers.

Effectiveness of RPC is confirmed by data on its use in responsible structures. RPC has been used in the construction of bridges, as well as other structures. Due to its increased strength, durability and radiation resistance, RPC can be used as a reliable material for containers of radioactive wastes from nuclear power plants. It is also used for thermal protection of buildings, as it provides better fire and heat resistance than ordinary high-strength concrete.

Research results show that RPC allows us to expand the possibilities of using concrete in new thin-walled structures, construction of which was previously impossible. Despite the fact that RPC production costs are generally higher than for conventional concrete, there are still some economic advantages for using RPC. Using dispersed reinforcement with short steel fibers yields enables us to minimize or even avoid the use of reinforcing bars. Due to the ultra-high mechanical properties of RPC, the concrete elements thickness can be reduced, which leads to savings in materials and the structure constructions cost.

RPC contains no coarse aggregate. Instead, fine powders such as quartz sand and milled quartz with a particle size of 0.045 to 0.6 mm are used. The term "reactive powder" reflects the fact that the dispersed components in RPC undergo chemical transformations at curing.

The main principle of improving concrete properties, implemented in RPC is to eliminate the drawbacks that are typical for ordinary concrete. From this viewpoint, the obtained ultra-high mechanical properties of RPC can be explained by the following features:

1. Increasing the RPC homogeneity by eliminating coarse aggregates. It is suggested that the maximum size of the components in RPC should be up to 0.6 mm;
2. Increasing the concrete density by optimizing the grain composition of the components mixture;
3. Forming a dense microstructure by heat processing after curing;
4. Improving the cementitious matrix properties by adding pozzolanic admixtures, such as microsilica and by reducing the water-binder ratio.

5.1 Influence of RPC Structural Features on its Strength

The main properties of RPC are caused by the lows of the structural theory of concrete [4]. According to Powers, compressive strength of the specimens of cement stone with various water-cement ratio that have hardened at normal temperature conditions, at different cement stone ages corresponds to the following empirical equation:

$$f_{cm} = AX^n \tag{5.1}$$

where X is the ratio of the cement gel volume to that of gel and capillary space;

A is a coefficient that characterizes the cement gel strength;

n is constant, which, depending on cement characteristics varies from 2.6 to 3.

Parameter X can be considered as the cement stone relative density. Representing Eq. (5.1) as a function of porosity P yields:

$$f_{cm} = A(1 - P)^n \tag{5.2}$$

When replacing porosity by relative density $d = 1 - P$, Eq. (5.2) takes the form:

$$f_{cm} = Ad^n \tag{5.3}$$

The value of parameter d is uniquely related to the water-cement ratio W/C:

$$d = \frac{1 + 0,25\alpha\rho_c}{1 + \rho_c.W/C}$$

(5.4)

where α is cement hydration degree;

ρ_c is cement density.

The first design dependences for high-strength concrete (HSC) strength were based on the general nonlinear nature of concrete strength dependence on C/W and extensive experimental data for $C/W \geq 2.5$:

$$f_{cm} = A_1 R_c (C/W + 0.5)$$

(5.5)

The value of coefficient A_1 depends on the initial materials quality.

Materials for concrete: A_1;

      high quality 0.43;

         ordinary 0.4;

      low quality 0.37.

For a more accurate consideration of the factors affecting the concrete strength, coefficient A can be expressed by the product $pA = A \cdot A_1 A_2 ... A_n$, де $A_1 ... A_n$, where $A_1 ... A_n$ are additional coefficients that take into account the dependence of concrete strength on temperature, curing time, admixtures, etc.

The possibilities of the design equations for concrete strength significantly increase when using the "modified C/W", $(C/W)_m$ taking into account the "cementing efficiency" coefficient or "cement equivalent" of the mineral admixture.

Using $(C/W)_m$ allows developing rather simple universal methods for calculating the composition of multicomponent concrete, based on the same physical preconditions.

The equation of concrete strength vs. $(C/W)_m$ has the following general form:

$$f_{cm} = pAf_{cem}((C/W)_m + b)$$

(5.6)

where pA_i is multiplicative coefficient $(pA_i = A_1, A_2 ... A_n)$, which takes into account a number of technological factors affecting the concrete strength at constant $(C/W)_m$, concrete age, temperature and humidity conditions of hardening, etc.

Using the modified C/W is rational, especially for design of concrete compositions with dispersed mineral admixtures. In concrete without mineral admixtures Eq. (5.6) is transformed into a usual form.

It is shown experimentally that at a given cement stone porosity the lower the pore size, the higher the strength (Fig. 5.1). The main assumption of modern structural theories is that concrete is a heterogeneous body. Presence of pores and cracks is an integral feature of the concrete structure. Capillary and other pores can be considered as cracks at a certain scale level. The cracks origin is usually associated with movement and modification of dislocations in the crystal lattice.

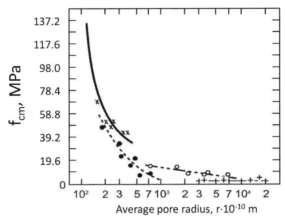

Figure 5.1. Relationship Between Compressive Strength f_{cm} and Average Pore Size of Cement Stone.

For heterogeneous material, volumetric deformations, temperature and humidity deformations of separate components, sedimentation phenomena, temperature and humidity gradients, osmotic phenomena, corrosion caused by the operating environment have a valuable effect on crack formation.

The concrete destruction process can be considered as the development of cracks, which usually occurs at the contact of the cement stone matrix and aggregate during concrete production and hardening up to their merging into through cracks. When the aggregate strength does not exceed the strength of the matrix, cracks can spread to the aggregates grains.

Studies carried out using an electronic scanning microscope and low-angle radiography have shown that the concrete failure process begins with formation and development of microcracks in hydrosilicates crystallites, cement hardening products, located near stress concentrators (pores and other technological defects). Cracks propagate from one grain to another until final macrodestruction becomes possible. At this stage free elastic energy is sufficient to form new surfaces of the main crack and also for the additional work required for plastic deformation and formation of stepped splitting surfaces. The crack propagation condition in polycrystalline materials is expressed by Griffiths-Orovan equation:

$$\sigma = \sqrt{Ev/d_{av}} = kd_{av}^{-1/2} \tag{5.7}$$

where σ is stress;

E is modulus of elasticity;

v is effective failure energy;

d_{av} is the average crystallite size;

$k = (Ev)^{1/2}$ is fracture toughness coefficient.

For porous bodies Eq. (5.7) should include additional coefficients that take into account the porosity and other defects. Experimentally confirmed dependence (5.7) (Fig. 5.2) showed that the disperse-crystallite structure of hydrosilicate

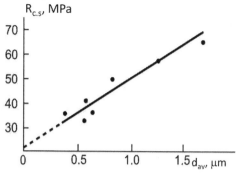

Figure 5.2. Dependence of Cement Stone Strength $R_{c.s.}$ on the Average Size of Crystallites d_{av}.

gel can be controlled by changing the curing conditions, introducing various modifying admixtures, etc. This structure has a significant effect on the cement stone strength.

It was assumed that the strength and other physical and mechanical properties of concrete depend on the ratio of the absolute volume, taken by the crystal growth per cement stone unit volume, to the absolute volume of gel. This ratio depends mainly on the cement mineralogical composition. The assumption considers no formation and development of separation microcracks and the effect of aggregates.

Analyzing experimental data, related to fracture microcracks formation, and comparing them with analysis of stress fields in concrete, has shown that the stress field, caused by the load, that interacts with that, caused by the material heterogeneity, resulting in local stress concentrations that lead to rupture cracks. Stresses in concentration zones are several times higher than those, caused by the load.

At repeated (cyclic) loading action microcracks appear in concrete long before failure which has so-called fatigue nature. This process is more intensive than in the case of static loading. Under static loads, local over stresses can relax due to the viscous properties of material, as well as stress and strain fields' redistribution.

Repeated high-frequency (vibration) loads reduce viscous bonds, increase the temperature in concrete microvolumes, which contribute to cracking. Concrete ability to resist fatigue failure characterizes its endurance (Fig. 5.3).

The safe stress, at which concrete can withstand repeated loads for an almost unlimited time is called the endurance limit σ_{end} and the relative endurance limit $K_{end} = \sigma_{end}/f_{cm}$ depends on the concrete structure, type and features of raw materials, curing conditions and concrete age. Depending on the type of aggregates after 1 million cycles K_{end} ranges from 0.7 to 0.38 and almost linearly related to the ratio of tensile or flexural and compressive strengths.

An increase in this ratio, which is characteristic for RPC, should increase the relative endurance limit.

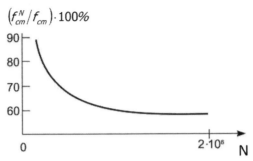

Figure 5.3. Dependence of Concrete Relative Endurance Limit on the Load Cycles Number N: f_{cm}^N, f_{cm} are respectively, Concrete Strength after N cycles and Initial Strength.

Influence of pulse and vibration loading is typical for bridges, road pavements, sleepers, bases under equipment, crane tracks, etc. Such loads are usually dynamic, because the load change duration from minimum to maximum values is measured in fractions of a second. Concrete failure under seismic impact has also a dynamic manner. Concrete resistance to seismic loading is characterized by seismic resistance coefficient, the ratio between the actual critical acceleration of the perturbing force, causing failure, to that of the strength limit. At other equal conditions, concrete with a high ratio between tensile and compressive strengths that is RPC also has the best resistance to seismic impact.

Along with repeated loading, under certain conditions concrete can undergo a single dynamic load. It is caused by explosive, shock and some other loads (Fig. 5.4).

The main indicator that determines concrete behavior under dynamic loading is dynamic hardening coefficient K_d, i.e., the ratio of the concrete strength under dynamic loading due to its static strength. Concrete dynamic strength is particularly sensitive to the presence of defects and microcracks in concrete, especially in the contact zone. As the concrete structure defectiveness increases

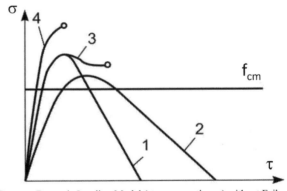

Figure 5.4. Concrete Dynamic Loading Model (stress σ vs. time τ) without Failure 1, 2 and with Failure 3, 4.

with increase in W/C, concrete dynamic strength decreases more than static strength.

The dynamic hardening coefficient depends on the loading duration τ:

$$K_d = a - b \lg \tau \tag{5.8}$$

where τ is the loading duration; a and b are empirical coefficients.

RPC structure peculiarities include a high cement stone content and high adhesion level to the admixture. Concrete strength at axial compression is determined by the ultimate tensile strength of the cement stone crystalline growth in the direction normal to the external forces action, without change in adhesion with aggregates. It can be described by the following equation:

$$f_{cm} = \frac{f_{c.s}}{1 - \dfrac{(\mu_c E_a - \mu_a E_{c.s})(1 - \Theta)q_a}{(E_{c.s}q_c + E_a q_a)\mu_c}} \tag{5.9}$$

where $f_{c.s}$ is the cement stone strength;

μ_c, μ_f are Poisson coefficients (transverse deformations coefficients) of cement stone and aggregates;

Θ is slip coefficient in case of partial lack of cement stone adhesion with aggregates

q_c and q_a are volumes of cement stone and aggregates per unit volume of concrete;

$E_{c.s}$ and E_a are modulus of elasticity of cement stone and aggregate.

As it follows from Eq. (5.9), depending on the aggregates elastic characteristics and their adhesion to the cement stone, the concrete strength may be higher or lower than that of cement stone. At maximum concrete strength adhesion of aggregates with cement stone is not broken ($\Theta = 0$). If the strength of aggregates is equal or higher than cement stone strength, it does not directly affect the strength of concrete.

To identify the relationship between concrete strength and the aggregate mechanical properties, the thermodynamic theory of solid bodies strength can be used. Elastic energy (W), spent for formation of new surfaces during fracture, is proportional to the square of the concrete strength:

$$W = f_{cm}^2 / 2 \tag{5.10}$$

Solving the differential equations within the limits of concrete two-component model components' volume concentrations change and performing certain transformation, an expression for concrete strength vs. mortar strength f_m and aggregate volume concentration V_a was obtained:

$$f_{cm} = f_m \left(1 - \frac{\alpha_{inh}}{2} K_a V_a\right) \tag{5.11}$$

where α_{inh} is coefficient that determines the effect of structural inhomogeneity of the matrix and the adhesion strength,

Table 5.1. Classification of Superplasticizers.

Name	SP nature	Action nature
NF	Based on sulfonated naphthalene-formaldehyde polycondensates	Electrostatic
MF	Based on sulfonated melaminoformaldehyde polycondensates	Electrostatic
LS	Based on sugar-free lignosulfonates	Electrostatic
P	Based on polycarboxylates and polyacrylates	Steric

K_a is coefficient depending on degree of concrete saturation by coarse aggregate.

Adding SP is currently a prerequisite for the production of high-quality RPC using modern technologies. The known SP are classified by their composition and mechanism of action (Table 5.1). The SP mechanism of action depends on a complex of physicochemical processes in the cement paste-admixture system. This mechanism is determined mainly by:

1. Adsorption of mono- or polymolecular SAS on the surface of mainly hydrated neoplasms;
2. Colloid-chemical phenomena at the phases border.

Addition of new generation SP on polycarboxylates basis provides an increase in a concrete mix cone slump from 3 cm to 25 cm and more. If concrete mixes with 25 cm and more, admixture of traditional SP quickly lose workability, mixes with polycarboxylates admixture are in a plastic state during 1.5 ... 2 h. High storage capacity of concrete mixtures with polycarboxylate and polyacrylate SP makes them especially attractive for monolithic construction and long-term transportation. They are successfully used at heat and moisture concrete processing in precast concrete industry.

5.2 Aggregates Grain Composition

For high-strength concrete, including RPC, the one of the most important technological tasks is optimization of aggregates grain composition. The first research on concrete aggregates grain composition design was aimed at ensuring the minimum voidness for mixtures of grains with different shapes and size [155–157]. At known values of aggregate bulk density $\rho_{b.ag}$ and grains density ρ_{ag} the calculated value of the its voidness V_n is:

$$V_n = 1 - \frac{\rho_b^m}{m_m}\left[V_1(1-V_{n_1}) + V_2(1-V_{n_2}) + ... + V_n(1-V_{n_n})\right], \qquad (5.12)$$

where: m_m and ρ_b^m are weights of the aggregates mixture and its bulk density;

$V_1, V_2 ... V_n$ are bulk volumes of the mixed fractions;

$V_{n_1}, V_{n_2} ... V_{n_n}$ are the voidness of the mixed aggregates fractions.

When mixing aggregates, the voidness value can be calculated as:

$$V_n = \left(1 - \frac{\rho_{b.ag}}{\rho_{ag}}\right) \qquad (5.13)$$

Analyzing the geometric structure of volume filled by grains of bulk material, the values for calculating the voidness of grains with different geometric shapes most and least dense stacking were proposed (Table 5.2). In practice, the most and least dense grain stacking is not real. The values of voidness increase for the higher angularity of grains, especially when using grains of an elongated shape. Two approaches have been formed to provide dense mixtures of aggregate grains: the choice of intermittent or continuous composition.

As known, at cubic placement of small and large balls with diameters d and D, respectively, to ensure the most dense stacking, the following conditions should be satisfied:

$$d = D\sqrt{2} - D = 0.41D \qquad (5.14)$$

For tetrahedral casting

$$d = \frac{2}{3}D\sqrt{3} - D = 0.155D \qquad (5.15)$$

Cubic packaging of balls has a voidness of 47.6%, tetrahedral—26%. The optimal ratios of diameters between the smallest grain fraction to that of the largest fraction range from 0.07 to 0.04.

Table 5.2. Voidness of Bulk Materials, Depending on Grains Shape.

Grains shape	Voidness, %		
	At most dense stacking	At least dense stacking	Average value
Cubic	0	87.1	43.55
Octahedric	12.1	83.9	48.05
Dodecahedrons	14.1	60.7	37.4
Icosahedrons	10.3	59.9	35.1
Spheric	26.2	47.6	36.9

The required amount of the n-th fraction of multifraction aggregate can be calculated as follows:

$$Q_n = \rho(1 - \varphi_1')(1 - \varphi_2')...(1 - \varphi_{n-1}')\varphi_n'' \qquad (5.16)$$

where ρ is material density;

φ_1', φ_2', φ_n' are filling coefficients, showing the change in the volume of voids of a coarse fraction when sequentially filled by smaller grains.

In practice, the values of φ_2' ... φ_n' can be taken as equal to 0.2. Then:

$$Q_n = \rho(1 - \varphi_1')(0.8^{n-2})\varphi_n'' \qquad (5.17)$$

where n is the number of fractions;

φ_n'' is the filling factor, which shows the required amount of the last fraction. For granite, basalt, limestone $\varphi_n'' = 0.307... 0.351$.

The two approaches have been formed to provide dense mixtures of aggregate grains: the choice of intermittent or continuous composition.

Although intermittent grain composition provides less void mixture of grains most researchers prefer continuous grain composition of concrete mixers. This is due to the need for the same concrete mixtures workability in the latter case a smaller volume of fine fractions and, accordingly, the cement consumption for grains coating. In addition, mixtures with a continuous grain composition are less prone to segregation.

To select a continuous grain composition of aggregates, various "ideal" sifting curves were proposed.

The most popular are the Fuller, Bolomei and Hummel curves, which are expressed by rather similar equations:

Equation	Author	
$y = 100\sqrt{\dfrac{d}{D}}$	Fuller	(5.18)
$y = A + \left(100 - A\sqrt{\dfrac{d}{D}}\right)$	Bolomei	(5.19)
$y = 100\left(\dfrac{d}{D}\right)^{n}$	Hummel	(5.20)

In Eqs. (5.18...5.20) y is passage of grains through sieve with a size d, %;

D is the maximum size of the aggregate;

A is a coefficient; when using crushed stone and sand—A = 10 for stiff and 12 for plastic concrete mixtures;

n is index an which can range from 0.1 to 1.

A method for obtaining curves of dense mixes, used for selecting aggregate content of asphalt concrete, was proposed by N. N. Ivanov. It is based on the assumption that the volumes ratio of each subsequent fraction to the previous one with maximum size that is 2 times higher (so-called coincidence coefficient), equal to k = 0.81. When calculated by Fuller's equation, this ratio (at equal densities of all fractions) is equal to $k = 0.707$, according to Hummel's equation at $n = 0.3$ $k = 0.812$. Rather dense mixtures can be obtained when the coincidence coefficient is in the range of 0.65 ... 0.8 (Fig. 5.5).

Assuming that the content of the first fraction in % is equal to a, then the content of the second will be $a_1 k$, the third—$a_2 k$, etc. The amount of the last fraction should be equal to $a_{n-1} k$. The sum of volumes of all fractions can is:

$$a(1 + \kappa + \kappa^2 + \ldots + \kappa^{n-1}) = 100\% \tag{5.21}$$

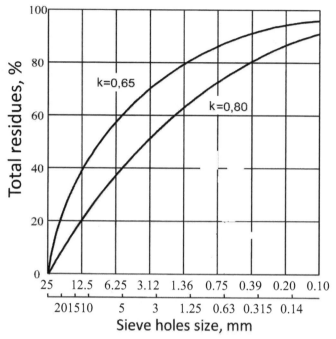

Figure 5.5. Curves of Dense Mineral Mixtures at Different Values of Coincidence Coefficient (k).

Therefore, the content of the first fraction (i.e., the partial residue on the corresponding sieve) is:

$$a_1 = \frac{1-\kappa}{1-\kappa^n} \cdot 100 \qquad (5.22)$$

For cement concretes the use of ideal curves for dense mixtures of aggregates is rational only in some cases, for example, for pressed or vibropressed concrete, which are made, from bulk mixtures.

For real aggregates, the grain composition always deviates from the ideal curve, so that the standards that define the requirements for aggregates indicate the recommended grain compositions range, beyond which cement overconsumption is significant. Aggregates voidness varies from 20 to 50% and is regulated by their fractionation followed by mixing.

An integral characteristic of fineness that is widely used to assess the quality of sand as a concrete aggregate, is the fineness modulus

$$M_f = \frac{A}{100} \qquad (5.23)$$

where A is the sum of the total residue on the control sieves, %:

$$A = A_{2.5} + A_{1.25} + A_{0.63} + A_{0.315} + A_{0.16} \qquad (5.24)$$

The total residues on each of the sieves are found by summing the partial residues on this sieve and sieves with larger holes.

The fineness modulus of sand characterizes the area above the integral sieving curve (Fig. 5.6). A disadvantage of this indicator is that it ambiguously characterizes the grain composition: the same value of M_f can correspond to different sieving curves (Fig. 5.6). To limit the area of possible variation of the sieving curves for assessing the sand quality total residue on the sieve No. 063 is indicated additionally to fineness modulus.

For plastic concrete mixtures, a significant effect on the optimal ratio of aggregate fractions have the cement paste layer thickness and consistency, which vary depending on concrete mixture and hardened concrete properties. From the viewpoint of cement consumption minimization, while providing a set of necessary concrete properties it is important that the aggregate composition provides the minimum possible voidness at the lowest total surface area. In RPC introduction of disperse admixtures-mineral admixtures helps minimize the voidness of aggregates and the thickness of the cement paste film.

The aggregate specific surface (U) can be obtained by the average grain size d_{av}:

$$U = \frac{N}{d_{av}} \qquad (5.25)$$

where N is coefficient depending on the grain shape and its surface relief.

Ideally, for a polished ball $N = 6$, for real bulk materials, this ratio is much higher. Table 1.14 presents the specific surface area values (U, cm^2/cm^3) and N for different fractions of crushed stone and gravel.

Various empirical formulas can be used to determine the calculated specific surface area of aggregates. For example, for quartz sand and crashed granite stone, the specific surface area can be calculated as follows:

$$U_s = \frac{0.025}{\rho_n}(a_1 + 2a_2 + 4a_3 + 7.4a_4 + 15a_5 + 110a_6) \qquad (5.26)$$

where $a_1...a_6$ are the contents of sand fractions at standard sieving, %;

Adjustment of aggregate parameters by mixing, for example, two sands can be performed using the following formula:

$$n = \frac{P_1 - P}{P_1 - P_2} \qquad (5.27)$$

where P is the required value of the adjusted parameter (fineness modulus, specific surface area, grain content of a certain fraction);

P_1 and P_2 are the value of the adjusted parameter in aggregate, according to the higher and lower values respectively;

n is the volume fraction of aggregate with a lower value of P in the sum of the mixed aggregates volumes.

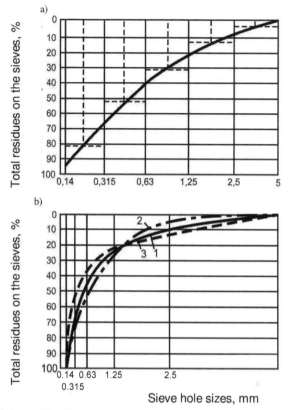

Figure 5.6. Integral Sand Sieving Curves: a - in the semi-logarithmic coordinate system; b - variants of sieving curves (1–3) sands with equal fineness modulus.

5.3 Activity of Ash-metakaolin Compositions and their Effect on Hydration and Structure Formation of Cement Stone

For wider use in the construction of RPC the relevant task is to replace silica fume with a more accessible active mineral admixtures. In this regard, the use of fly ash, metakaolin admixture and their compositions is of practical interest.

The big advantage of fly ash as cement concrete admixture is its low water demand. Adding ash into the concrete mixes as a rule improves its workability.

However, the relatively low pozzolanic activity of ash does not allow it to ensure a high degree of hydration of the cement matrix especially in the curly stages of its hardening.

High pozzolanic activity has the product of kaolin dehydration-metakaolin (MK), however, it significantly increases the water demand of concrete mixtures. It can be assumed that composite ash and metakaolin admixture should have, on the one hand, increased pozzolanic activity and surface energy, and on the other hand, moderate water demand and ensure long-term strength growth, of concrete. This hypothesis is confirmed by the results of the studies presented in this book.

Autohesion. The structure-forming role of mineral admixtures in cement systems is, to a certain extent, related to the surface energy magnitude, which in turn is evident in particles adhesion or autohesion strength. The main factors determining the powders autohesion magnitude (p_a) are the particles size and surface tension, which follow from the equation

$$p_a = 2\pi \frac{d_1 d_2}{d_1 + d_2}(\sigma_1 - \sigma_2) \tag{5.28}$$

where d_1 and d_2 are particles size;

σ_1 is surface tension of solid particles at the border with a dispersed medium;

σ_2 - same on the solid particles contact border.

The influence of the studied powder materials type and ratio on adhesion strength is shown in Table 5.3.

Autohesion of powders was determined experimental by the method of simultaneous separation. The autohesion valul in a row MK – Portland cement – fly ash decreases. Along with the experimental data, Table 5.3 shows the mixtures autohesion P_a calculated using the admixturely formula:

$$P_a = P_{a_1}\varphi_1 + P_{a_2}\varphi_2 + ... + P_{a_n}\varphi_n \tag{5.29}$$

where $P_{a_1}, P_{a_2},...P_{a_n}$ are the adhesion strength values of powder particles forming the mixture;

$\varphi_1, \varphi_2,...\varphi_n$ is the powders content ratio.

Analysis of the obtained data shows that the additivity rule of mixtures autohesion is valid mainly for a cement-ash system, which has a relatively small difference in the dissimilar particles size. At big difference in the contacting powders particle sizes, characteristic of MK-fly ash and MK-fly ash-Portland cement mixtures, a synergistic autohesion effect is observed.

Adding up to 20% MK to ash increases autohesion and brings it to the value of P_a for pure metakaolin. An increase in MK content to 40% increases autohesion by 15–20%, a further increase in the MK proportion in the mixture leads to a gradual decrease in autohesion. The decrease in P_a with an increase in the big particles content in the mixture can be explained by the reorganization of the powder structure with a decrease in the fracture surface. The change in autohesion in cement-MK and cement-ash-MK systems is of approximately the same nature. In the latter case, cement and ash are sources of relatively big and MK - of super-thin particles.

Wetting heat. A certain idea of the powders surface energy, energetics of their interaction with the liquid phase, and the adhesive ability allow wetting heat measurements. The values of wetting heat determined using on adiabatic calorimeter [29] for the studied materials are given in Fig. 5.10.

From the wetting heat expression (Q) follows:

Table 5.3. The Effect of Powder Components Composition on Autogesion.

Mixtures composition, %		Autohesion (experimental/ calculated), Pa	Mixtures composition, %			Autohesion (experimental/ calculated), Pa
Metakaolin MK	Fly ash A		Metakaolin MK	Fly ash A	Portland cement	
0	100	17.8	5	15	80	26.5/24.1
5	95	20/18.6	10	10	80	29.8/24.9
10	90	26.7/19.8	15	5	80	30.2/25.7
15	85	28.8/20.3	5	25	70	24.7/23.4
20	80	31.6/21.1	10	20	70	28.1/21.8
30	70	36.5/22.8	5	35	60	23.8/22.7
40	60	38.9/24.4	10	30	60	27.3/23.5
50	50	39.9/26.1	15	25	60	29.1/24.4
60	40	39.7/27.8	5	45	50	23/1/22
70	30	38.5/29.4	10	40	50	26.8/22.9
80	20	37.6/31.1	20	30	50	31.9/24.5
90	10	37.1/32.7	-	30	70	21.4/22.6
100	0	34.4	-	50	50	20.5/21.2
			10	-	90	25.1/25.6
			20	-	80	33.4/26.6
			30	-	70	37.1/27.6

$$Q = S(\varepsilon_{s-g} - \varepsilon_{s-l}) \qquad (5.30)$$

where ε_{s-g} and ε_{s-l} are total surface energy on solid - gas and solid-liquid interface accordingly;

S is the powder specific surface.

The obtained wetting heat values for MK and fly ash are characteristic for dispersed hydrophilic materials. According to the Laplace equation, interfacial tension surfaces characterizing the values of free surface energy, σ, are related as follows:

$$\sigma_{s-g} = \sigma_{s-l} + \sigma_{l-g} \cos \alpha \qquad (5.31)$$

where α is the wetting angle.

For hydrophilic materials, assuming complete wetting, i.e., cos α = 1, Eq. (5.31) takes the form

$$\sigma_{s-g} = \sigma_{s-l} + \sigma_{l-g} \qquad (5.32)$$

Passing from free energy σ to total surface energy ε it can be assumed that

$$\varepsilon_{s-g} - \varepsilon_{s-l} = \varepsilon_{l-g} \qquad (5.33)$$

And from Eq. (5.30...5.33) yields that

$$S = Q/\varepsilon_{l-g} \qquad (5.34)$$

Using the reference value of the total surface energy ε_{l-g} = 0.117 J/m^2 for water, the calculated values of S can be obtained. For two metakaolin samples differing in dispersion they turned out to be equal to 22,1 and 28,3 m²/g, ash 1.45 and 1.54 m²/g.

Estimating the obtained results for S, it can be noted that they are significantly higher than the experimental ones, obtained by the air penetration method. Figure 5.7 shows the experimental values of wetting heat in water Q for mixtures of metakaolin and ash. This data indicates that the addition of MK to fly ash increases the total surface energy of the mixture not according to a linear law of mixtures, but in accordance with a following exponential dependence

$$y = Ax_1^\beta \ (\beta<1) \tag{5.35}$$

where A and b are constants,

$$x_1 = \upsilon_{MK}/(\upsilon_{MK}+\upsilon_a),$$

where υ_{MK} and υ_a are volume concentrations of MK and ash accordingly.

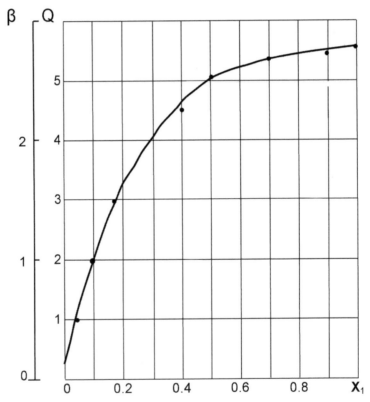

Figure 5.7. Influence of MK Volume Concentration of in a Mixture of MK and Ash on the Wetting Heat Q and the Coefficient β.

Pozzolanic activity. Pozzolanic activity and kinetics of lime absorption from a saturated solution were determined for various compositions MK and fly ash. When studying the kinetics of CaO absorption by active mineral materials, pre-heated and temperature-controlled at 80°C saturated lime solution and the investigated powders were mixed, and after a certain time the concentration of $Ca(OH)_2$ was measured by titration. The results of the experiments are given in Table 5.4 and Fig. 5.8. As it follows from the above data, pozzolanic activity of MK, is 3.5 ... 4.5 times higher than that of fly ash. Milling ash to $S = 420 ... 450\,m^2/kg$ increases its activity only 1.5 times, while mixing with 20 ... 30% MK allows to increase the lime absorption by almost 2 times.

Following the graphs representing kinetics of lime absorption by MK, fly ash and their mixtures (Fig. 5.8), two periods are evident: the first, corresponding to a sharp initial change in the concentration of $Ca(OH)_2$, and the second, characterized by a lower process speed. The first period of duration in MK and fly ash-metakaolin mixtures is about 3 ... 4 times shorter than in fly ash.

Separation of lime absorption by colloidal silica into two periods was previously noted in the work of P.E. Holsted and S.D. Lawrence [196]. The first period was interpreted as adsorption of $Ca(OH)_2$ on SiO_2 particles. The amount of adsorbed $Ca(OH)_2$ in a rough approximation is equivalent to a layer of Ca^{2+} ions on the surface of SiO_2 particles with a thickness of one Ca^{2+} ion. A continuing decrease in the concentration of $Ca(OH)_2$ is due to precipitation from a calcium hydrosilicate solution, formed as a result of a chemical reaction.

Pozzolanic reaction of ash also begins with the adsorption of calcium hydroxide on its surface. Under normal conditions, this film on ash particles

Table 5.4. Pozzolanic Activity of MK, Fly Ash and their Mixtures.

Material	Absorption of CaO, mg/g	Material	Absorption of CaO, mg/g
Metakaolin MK	155.4	MK 10% A 90%	46.5
		MK 20% A 80%	60.1
		MK30% A 70%	70.3
Fly ash (A)	35.5	MK 40% A 60%	81.8
		MK 50% A 50%	96.6
		MK 60% A 40%	107.4
		MK 70% A_M 30%	120.4
Fly ash(A_M) milled to S = 420...450 m^2/kg	57.3	MK 80% A_M 20%	132.8
		MK 90% A_M 10%	141.9

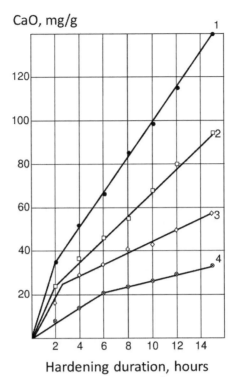

Figure 5.8. Kinetics of CaO Absorption by Metakaolin (MK) and Fly Ash (A):
1 - MK; 2 - MK - 50%, A - 50%; 3 - MK - 20%, A - 80%; 4 - A;

is formed in concrete within 24 hours, independent of the aggregate type. Thin boundary layers that arise in this case serve as conductors of calcium ions, under the influence of which a gradual erosion of the ash particles surface develops. Recesses are formed in it where the products of the pozzolanic reaction settle. In accordance with this mechanism, active centers exist on the surface ash. Particles of Ca $(OH)_2$ initially diffuse to these centers. Subsequently, deposition $Ca(OH)_2$ begins at the active centers and a spherical film of calcium hydroxide appears. Chemical interaction between the particles of $Ca(OH)_2$ and active glas phase completes the pozzolanic reaction of ash.

Addition of compositional MK significantly accelerates the binding of $Ca(OH)_2$ released during the hydrolysis of alite and the formation of low-basic hydrosilicates. Almost complete absorption of $Ca(OH)_2$ and, accordingly, the formation of hydrated neoplasms occurs when the content of ash-metakaolin admixture (AMK), including 20 ... 30% of MK, up to 40% by weight.

Accelerating the binding of $Ca(OH)_2$, MK promotes a more complete hydrolysis of clinker minerals and an increase in the degree of cement hydration. As can be seen from Fig. 5.9, the introduction of 30% MK brings the degree of cement hydration in 28 days to 0.8, while already at 3 days of age it is 0.65, i.e., 25% higher than that of cement without MK. The hydration-accelerating effect of admixture containing only 30% MK is quite close to the effect of pure MK. The ash component of the compositional admixture also contributes to the accelerating hydration effect.

In presence of ash, hydration of alite is accelerated after the induction period [197]. This is due to the chemisorption of Ca^{2+} ions on the ash surface and to the crystallization of the CSH phase on them. According to Uhikawa, under the same conditions, the degree of C_3S hydration reaches 35% in a day, and in a mixture with ash (specific surface 370 m^2/kg) it is already 45%. When assessing the effect of ash on the hydration of cement, one should take into account the content of alkalis and sulfates in them, which retard the hydration of C_3A and C_4AF.

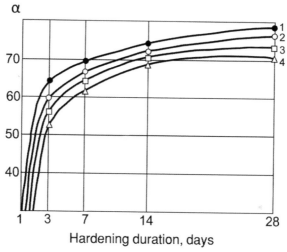

Figure 5.9. The effect of MK, Fly ash and their Mixtures on the Hydration Degree (α) of Portland Cement (PC):1 – 30% MK; 2 – 30% AMK (30% MK: 70% A);3 – 30% fly ash (A); 4 – PC.

A change in AMK proportion in the filled binder composition affects the hydration degree up to a certain limit, which is obviously related to the composite admixture and Portland cement composition. According to our experimental data, a noticeable increase in the cement hydration degree (α) occurs with an increase in the volumetric concentration of AMK (X_2) in the filled binder to 0.25. With a further increase in X_2, the value remains practically unchanged.

Rheological properties. To investigate the influence of the main composition factors on rheological properties of cement-water pastes with ash-metakaolin admixture (AMK), algorithmized experiments were performed in accordance with the standard B_4 plan [13]. The experiments planning conditions are given in Table 5.5.

As a result of experimental data processing, the following equation for effective viscosity in Pa·s is obtained.

Table 5.5. Experiments Planning Conditions for Studying the Effective Viscosity of cement-water pastes with ash-metakaolin admixture (AMK).

Factors			Variation levels			Variation interval
Name	Natural	Coded	−1	0	+1	
Volumetric concentration of MK in AMK	X_1	x_1	0	0.5	1	0.5
Volumetric concentration of AMK in filled binder	X_2	x_2	0.2	0.45	0.7	0.25
Volumetric concentration of filled AMK binder in water paste	X_3	x_3	0.6	0.65	0.7	0.05
Volumetric concentration of superplasticizer (SP) in water (W)	X_4	x_4	0	0.01	0.02	0.01

Note: in these and subsequent experiments were used Portland cement CEM-1, metakaolin, fly ash and naphthalene-formaldehyde superlplacticizer SP.

$$y_1 = 210.5 + 121.4\,x_1 + 54.6\,x_2 + 85.3\,x_3 - 105.3\,x_c - 28.5\,x_1x_4 - 21.7\,x_2x_4 +$$
$$55.3\,x_1^2 + 31.3x_2^2 + 15.4\,x_4^2 \qquad (5.36)$$

The obtained regression equation can be considered a mathematical model of viscosity for the investigated MK-fly ash-Portland cement–water-superplasticizer system under the assumed composition constraints. Analysis of the model shows that within the investigated compositions diapason, viscosity, varies over a wide range. By the influence on the value of viscosity (η) the factors can be ranked in the order $x_1 > x_c > x_3 > x_2$. When MK is added to ash admixture, viscosity of the paste increases unevenly as the volume concentration of MK in AMK (x_1) increases. If an increase in x_1 to 0.5 yields viscosity increase by 66 Pa·s, and if x_1 increases from 0.5 to 1 viscosity increases by 177 Pa·s. When $x_1 < 0.5$, it is possible to prevent an increase in pastes viscosity at superplasticizer concentration in water up to 0.01. Complete replacement of fly ash by MK is not compensated even at a double concentration of superplasticizer.

The possibilities to control viscosity of cement-water pastes filled with AMK can be analized by the viscosity isolines diagrams (Fig. 5.10). As it follows from diagrams, the main technological method for reducing the viscosity of cement-water systems containing MK and AMK is using the appropriate concentration of superplasticizer.

Analysis of the viscosity model enables us to get to another important conclusion: adding MK in composition with ash leads to a decrease in cement-water pastes viscosity in comparison with that of pastes containing pure MK.

A change in the rheological properties of filled with AMK admixture cement-water pastes is also evident in their normal consistency (Table 5.6). As it follows from Table 5.6, using AMK in combination with superplasticizer yields no significant increase in normal consistency and, consequently, in water demand of cement pastes. At the same time, when cement pastes are filled just by MK,

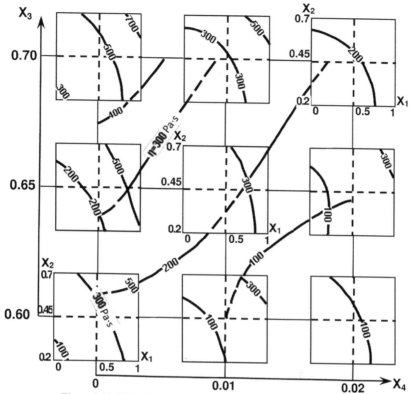

Figure 5.10. Viscosity Isolines for Cement-water Pastes with AMK.

Table 5.6. Normal Consistency of Filled Cement-water Pastes with AMK Admixtures.

Compositions of pastes*			
$X_1 = V_{MK}/(V_{MK} + V_a)$	$X_2 = (V_{MK} + V_a)/(V_{MK} + V_a + V_{sp})$	$X_4 = V_{sp}/(V_{sp} + V_w)$	N.C. , %
0	0	0	24.5
1	0.1	0	29.8
1	0.5	0	36.3
0	0.2	0	23.7
0	0.5	0	24.8
1	0.2	0.01	27.4
1	0.2	0.02	26.5
1	0.5	0.01	32.1
1	0.5	0.02	29.4
0.2	0.2	0	28.1
0.2	0.2	0.01	25.1
0.2	0.2	0.02	23.6
0.2	0.5	0.01	24.3
0.5	0.2	0.01	26.8
0.5	0.2	0.02	24.9
0.5	0.5	0.01	27.7
0.5	0.5	0.02	25.4

*The factor is taken constant $X_3 = 0.65$.

their normal consistency remains significantly higher even if the superplasticizer concentration is high.

Particularities of structure formation. Structure formation kinetics of normal consistency cement pastes, filled by AMK, was studied by measuring the plastic strength and ultrasonic waves' transmission velocity. A smooth increase in plastic strength over the first section of plastograms (Fig. 5.11) corresponds to the coagulation structure formation period. Particles of admixture form coagulation contacts with hydrated cement particles. For the globular-type structures formed in this process, the contact strength (R_{cont}) depends on a number of factors

$$R_{cont} = \gamma f(F_p, \varphi, S_p) \tag{5.37}$$

where γ is a chemical interaction constant;

F_p is resulting interaction force between the particles;

φ is the filling degree;

S_p is specific surface of particles involved in the interaction.

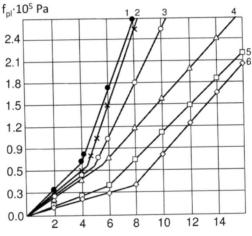

Figure 5.11. Kinetics of Cement-water Pastes Plastic Strength Change (f_{pl}):
$1 - X_1 = 1$; $X_2 = 0.2$; $X_4 = 0$; $2 - X_1 = 0.5$; $X_2 = 0.2$; $X_4 = 0$; $3 - X_1 = 0.2$; $X_2 = 0,5$; $X_4 = 0$; $4 - X_1 = 1$; $X_2 = 0.2$; $X_4 = 0$; $5 - X_1 = 0$; $X_2 = 0$; $X_4 = 0$; $6 - X_1 = 0$; $X_2 = 0.5$; $X_4 = 0$.

When adding both MK and AMK, an increase in the coagulation contacts strength can be expected, which is confirmed by plastograms (Fig. 5.11). It is typically that at approximately equal volume concentrations of metakaolin and fly ash—metakaolin admixture, the absolute values of cement pastes plastic strength and kinetics of its changes are quite close.

The first part of the plastograms approximately corresponds to the cement pastes setting time, and the inflection point—to the final of setting (Table 5.7).

Table 5.7. Setting Time of Cement-water Pastes with Mineral Admixtures.

Composition of pastes*			Setting time, h-min.	
X_1	X_2	X_4	initial	final
0	0	0	2–10	6–05
1	0.2	0	2–40	6–15
1	0.5	0	2–55	6–25
0	0.2	0	2–30	6–20
0	0.5	0	3–15	7–31
1	0.2	0.01	1–40	5–40
1	0.2	0.02	1–28	4–25
1	0.5	0,01	2–15	6–18
1	0.5	0.02	2–05	5–50
0.2	0.2	0	2–30	6–43
0.2	0.2	0.01	1–52	5–35
0.2	0.2	0.02	1–52	5–14
0.2	0.5	0.01	2–25	5–35
0.5	0.2	0.01	2–05	5-24
0.5	0.2	0.02	1–44	5–21
0.5	0.5	0.01	2–17	5–44
0.5	0.5	0.02	2–05	5–25

*–x_3= 0.

Adding ash in cement pastes slows down their setting, which can be explained by the stabilizing effect of ash on the aluminate phase hydration.

MK can both slow down and accelerate the setting and, in particular, its beginning. According to the physical mechanism, cement paste setting is caused by an increase in the hydration products volume and a corresponding decrease in the distance between the particles until cohesion forces proportional to the number of particles in contact with each other begin to arise. A decrease in the distance between hydrated particles and strengthening of the coagulation structure with addition of MK is typical when water demand of cement pastes is limited due to superplasticizer.

The second part of the plastograms corresponds to the coagulation strenthening period of coagulation and beginning of the crystallization structure formation. At this stage addition of MK and AMK in combination with superplasticizer causes an fast increase in plastic strength, which is consistent with theoretical assumptions about their influence on the crystallization. After about 2 ... 3 hours after the final of setting, the plastic strength of cement pastes filled with MK and AMK and addition of superplasticizer increases by almost 4 times and reaches 2.7 Pa. For other pastes that were studied, this value of plastic strength is achieved after a significantly longer period of time. It is typical that in cement-water pastes of normal consistency, a change in the MK volume concentration in the AMK composition from 1 to 0.2 has a low effect on kinetics of plastic strength growth. At low water demand of cement pastes with addition of aqueous superplasticizer, and at the same time at sufficient concentration of active admixture with colloidal dispersion so called "cramped"conditions are created. Under "cramped conditions", so close interparticle contact takes place that forces of a different nature lead to hardening structure formation.

Kinetics of the plastic strength growth is consistent with that of ultrasonic waves' velocity passing through hardening cement-water pastes.

The initial period of coagulation structure formation on the ultrasound velocity curves is characterized by a horizontal part (Fig. 5.12), the length of which coincides well with beginning of setting. Determining the beginning of setting by the plastic strength value is difficult due to its lower sensitivity to initial contacts between particles. The structure formation rate can be related to the ultrasound velocity curves inclination angle, as well as the plastic strength, to the abscissa axis.

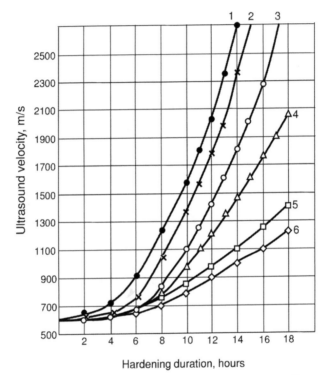

Figure 5.12. Kinetics of Ultrasound Velocity through Cement-water Pastes:
$1 - X_1 = 1; X_2 = 0.2; X_4 = 0.02; 2 - X_1 = 0; X_2 = 0.2; X_4 = 0.02; 3 - X_1 = 0.2; X_2 = 0.5; X_4 = 0.02; 4 - X_1 = 1; X_2 = 0.2; X_4 = 0; 5 - X_1 = 0; X_2 = 0; X_4 = 0; 6 - X_1 = 0; X_2 = 0.5; X_4 = 0.$

Strength of cement stone with AMK admixture. The active influence of ash-metakaolin admixture(AMK) on hydration and structure formation of cement-water pastes enables it to suggest its natural role in the synthesis of cement stone strength. To estimate the effect of AMK composition and the degree of cement pastes filling on strength of filled cement stone, algorithmic experiments were performed using Portland cement PC-1 and PC-2 with 28-days compressive strength 50 and 40 MPa, metakaolin and fly ash.

The effect of three factors on compressive strength of cement stone at 28 days was investigated: volumetric part of MK in AMK composition (X_1), volumetric part of AMK in the filled binder composition (X_2) and volumetric part of superplasticizer in the water (X_4). The planning conditions for these factors are given in Table 5.5. The experiments were performed in accordance with the three-factor B_3 plan. Statistical processing of the obtained results enabled to obtain strength regression equations adequate for 95% confidence probability for binders based on PC-1 (Y_2) and PC-2 (Y_3):

$$Y_2 = 56.5 - 8.5x_1 - 12.4x_1^2 - 5.5x_2 - 3.6\,x_2^2 + 9.2\,x_4 -$$
$$-4.3\,x_4^2 - 2.2\,x_1\,x_2 + 4.1\,x_1\,x_4 + 2.5\,x_2\,x_4 \qquad (5.38)$$

$$Y_3 = 45.6 - 6.7x_1 - 10.6x_1^2 - 3.9x_2 - 2.8\,x_2^2 + 7.4\,x_4 -$$
$$-3.6\,x_4^2 - 1.8\,x_1x_2 + 3.4\,x_1x_4 + 1.9\,x_2x_4 \qquad (5.39)$$

As follows from the models analysis, activity of cements can be substantially increased by filling them by AMK and using water solution of superplasticizer.

Analysis of the obtained equations demonstrates the nonlinear effect of all the investigated factors on the output parameter. The studied factors ranking according to their effect on strength yields:

$$X_1 > X_4 > X_2$$

The most significant interaction effects between factors X_1 and X_4, X_2 and X_4 show that an increase in the MK content in AMK composition and AMK binder filling degree in order to increase its activity requires simultaneous increase in superplasticizer concentration. At the same time, as the optimum values of X_1 and X_2 increase, and despite the maximum possible concentration of superplasticizer, the strength of cements filled by AMK decreases monotonously, which can be explained, first of all, by an uncompensated increase in binders water demand.

5.4 Influence of Ash and Slag Admixtures on Reactive Powder Concrete Strength

Fly ash of Burshtyn TPP (A) (specific surface area $S = 2527$ cm²/g), metakaolin (MK) "Western Kaolin Company LLC" ($S = 10125$ cm²/g), blast furnace granulated slag (BFGS) ($S = 2725$ cm²/g) were used as mineral admixtures in the experimental studies. For comparison reactive-powder concrete on silica fume SikaFume-HR TU (MK Sika) ($S = 23158$ cm²/g) was also used. During the experiment it was determined that the water demand (W/C) of the concrete mixture required to achieve the necessary fluidity of the mixture. The fluidity was controlled by measuring the flow diameter using a Suttard viscometer (25… 30 cm). Cubic specimens of $10 \times 10 \times 10$ cm and beams of $4 \times 4 \times 16$ cm were produced. Concrete compositions are given in Table 5.8, Mapei Dynamon SP3 polyacrylate superplasticizer was used at all points of the experiment.

The compressive and flexural strength at 1, 7 and 28 days of hardening under normal conditions were determined. For selecting the reactive powder concrete effective compositions C_f coefficient was calculated. This characterizes the specific consumption of cement per compressive strength unit. This coefficient was proposed to determine the effectiveness of reactive powder concrete compositions.

$$C_f = \frac{C}{f_c}, \text{kg/MPa} \tag{5.40}$$

where C is the cement content per 1 m³ of concrete, kg;

f_c is compressive strength of concrete, MPa.

The results of the experiment are given in Table 5.8 and in Fig. 5.13–5.15.

As it follows from the results (Table 5.8), RPC for a given composition, produced from high–fluidity mixtures with a maximum content of microsilica SikaFume-HR TU of 360 kg/m³ after 28 days of normal curing reaches a compressive strength of 162.4 MPa, which corresponds with available data. Partial or complete replacement of microsilica of this type by other investigated admixtures enables it to produce concrete with strength values of 104 – 160 MPa Figs. (5.13, 5.14,5.15). Such strength values at high concrete mixture fluidity indicate that the resulting material should be characterized by high technological and performance characteristics under the different loads that occur during the lifetime of various types of structures.

The flexural strength for all of the studied series of specimens ranged from 18.3 to 32.2 MPa. Such rather high strength values are important for using these materials in structures that have high requirements for dynamic loads. It should be noted that RPC has rather high ratio of flexural to compressive strength ($f_{cm}/f_{c.tf}$) = 4.0 – 5.0. As known, this indicator characterizes the concrete resistance to cracks formation and its deformability. For ordinary concrete the ratio $f_{cm}/f_{c.tf}$ is usually in a range of 7 – 8, for fine-grained concrete it is slightly below 5 – 6.

A significant amount of calcium hydroxide is released in RPC cement stone due to high cement consumption. It actively interacts with such mineral admixtures as ash or slag forming insoluble compounds. Fly ash due to its rounded grains shape has some plasticizing effect, which enables it to maintain low W/C values and a dense cement stone structure even at high consumption.

The maximum increase in strength was achieved by using the highly active admixtures of microsilica and metakaolin. SikaFume-HR TU micro-silica proved to be the most effective (maximum compressive strength, with a maximum consumption of 360 kg/m³ – 162.4 MPa), addition of metakaolin (133.4 MPa) showed a slightly lower efficiency (Fig. 5.15).

It should be noted that the maximum efficiency of metakaolin addition is observed at a consumption of about 10% by cement weight, this result confirms the data obtained by us earlier in the study on high-strength concrete [25]. A further increase in the amount of metakaolin leads to an increase in the concrete mixture water demand and the concrete strength decreases (Fig. 5.3).

Table 5.8. Influence of Various Mineral Admixtures on Compressive and Flexural Strength of Reaction-powder Concrete.

No.	Cement kg/m³	Sand fraction 0.16…1.25 mm, kg/m³	Mineral admixture, kg/m³	Water demand, l/m³	W/C	Flow, cm	Concrete strength					
							f_{ctf}^1, MPa	f_{cm}^1, MPa	f_{ctf}^7, MPa	f_{cm}^7, MPa	f_{ctf}^{28}, MPa	f_{cm}^{28}, MPa
						Microsilica SikaFume						
1	1080	1200	120	240	0.22	30	10.1	47	25.6	95	29.7	143
2	840	1200	360	240	0.29	12	12.7	43,3	28.2	125.5	32.2	162.4
						Blast furnace granulated slag						
3	1080	1200	120	240	0.22	15	6.19	36.1	20.83	82.4	24.9	113.8
4	840	1200	360	240	0,29	25	5,83	23,3	15,66	80,1	18,8	118,0
						Ash (A) (S = 2527 cm²/g)						
5	1080	1200	120	240	0.22	30	6.46	31.5	19.22	70.3	23.1	100.9
6	840	1200	360	240	0.29	30	4.76	22.8	15.22	68.5	18.3	114.5
						Metakaolin (MK)(S = 10 125 cm²/g)						
7	1080	1200	120	240	0.22	30	9.8	48.1	22.7	103.1	26.6	133.4
8	840	1200	360	270	0.33	10	7.8	27.2	17.2	78.5	20.1	102.4

Table 5.9. Influence of Various Mineral Admixtures on the Strength Efficiency Criteria.

No. (Table 5.8)	Concrete strength					
	$f_{cm}^1/f_{c.tf}^1$	$f_{cm}^7/f_{c.tf}^7$	$f_{cm}^{28}/f_{c.tf}^{28}$	C/f_{cm}^1, kg/MPa	C/f_{cm}^7, kg/MPa	C/f_{cm}^{28}, kg/MPa
Microsilica SikaFume-HR/TU						
1	2.7	3.7	4.8	40.0	11.4	7.6
2	2.6	4.5	5.0	25.2	6.7	5.2
Blast furnace granulated slag						
3	5.6	4.0	4.6	20.8	11.7	8.4
4	4.3	5.1	6.3	22.5	9.3	6.3
Ash (A)						
5	4.9	3.7	4.4	23.2	13.7	9.5
6	4.8	4.5	6.3	22.8	10.9	6.5
Metakaolin						
7	4.9	4.5	5.0	22.5	10.5	8.1
8	3.5	4.6	5.1	30.9	10.7	8.2

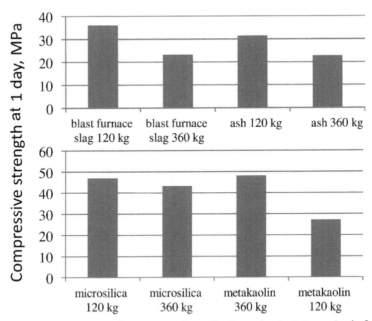

Figure 5.13. The Effect of Fine Mineral Admixtures on RPC Strength at 1 Day.

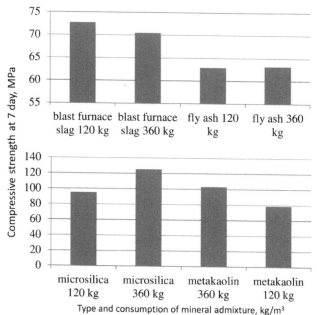

Figure 5.14. The Effect of Fine Mineral Admixtureson RPC Strength at 7 Days.

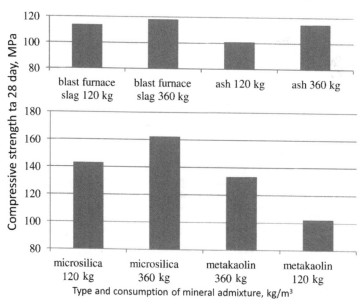

Figure 5.15. The Effect of Fine Mineral Admixtureson RPC Strength at 28Days.

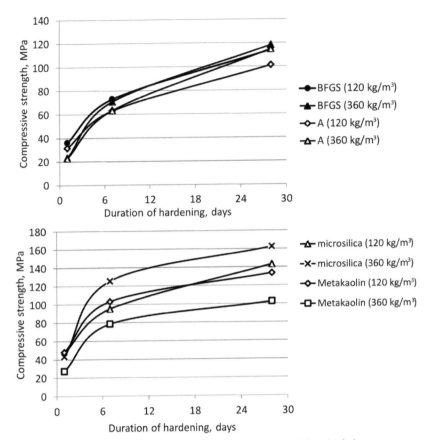

Figure 5.16. Kinetics of RPC Strength Increase for Various Mineral Admixtures.

Kinetics of RPC strength growth (Fig. 5.16) mainly corresponds to the known data. Series containing SikaFume-HR TU microsilica (consumption 360 kg/m³) and metakaolin (consumption 120 kg/m³) exhibited the maximum curing speed. All series, characterized by maximum strength, at the first day reached strength in the range from 40 to 48 MPa. This strength can be considered sufficient for high-speed construction.

The nature of different mineral admixtures effect on RPC flexural strength (Figs. 5.17–5.19) is almost the same like on compressive strength. Composites containing SikaFume-HR TU microsilica have the maximum $f_{c.tf}$ (29 – 32 MPa), for metakaolin and ash it is slightly lower (26 – 27 MPa).

The most effective compositions from the viewpoint of strength per kilogram of cement C_f were compositions containing high amounts of microsilica, fly ash, and ground slag (Table 5.9).

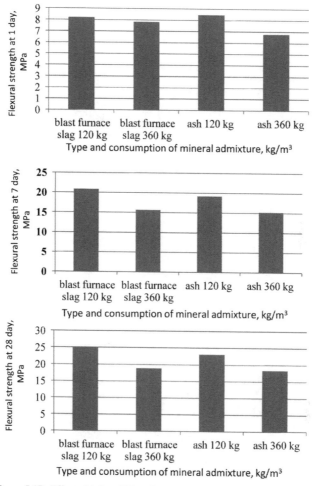

Figure 5.17. Effect of Ash and Blast Furnace Slag on RPC Flexural Strength.

Considering the cost, using admixtures such as fly ash, and ground slag may be the most economically justified. Given the obtained effectiveness for RPC mineral admixtures of different activities, a series of experiments were carried out to investigate their joint effect on concrete strength properties. During this investigation, the reactive powder concrete composition was constant: cement consumption – 840 kg/m³, mineral admixtures – 360 kg/m³, sand fraction 0.16… 1.25 mm – 1200 kg/m³, polyacrylate type superplasticizer Dynamon SP-3 – 2% by weight of the binder. Water demand was determined from the condition of providing a given flow of 25 – 30 cm on the Suttard viscometer. Mineral admixtures fly ash (A) and blast furnace granulated slag (BFGS) were used.

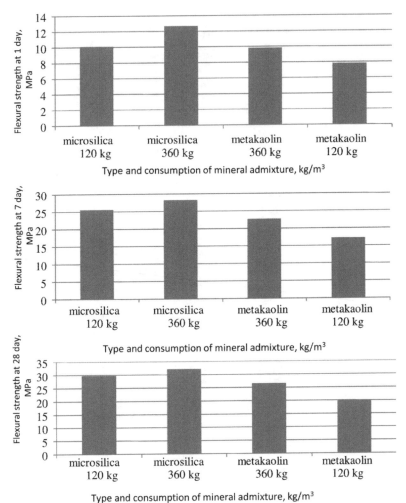

Figure 5.18. The Effect of Active Mineral Admixtures on RPC Flexural Strength.

The addition of metakaolin (MK)was in amount of 5, 10 and 15% by cement weight. The obtained results (Table 5.10, Fig. 5.20–5.23) indicate that the when using metakaolin admixtures concrete compressive strength increases to 122 – 126 MPa. However, an increase in metakaolin content of more than 10% by cement weight is inefficient due to a significant increase in the mixture viscosity and water demand. The most effective admixture in combination with metakaolin is fly ash—as it yields the maximum concrete strength. The positive effect of ash on the concrete mixtures fluidity compensates to some extent the increase in viscosity that is inherent for metakaolin. Effectiveness of the investigated mineral admixtures complexes from the viewpoint of increasing the flexural strength is similar. The maximum values of $f_{c.tf}$ (23 – 25 MPa) were achieved by combining metakaolin by the amount of 10% by cement weight with fly ash.

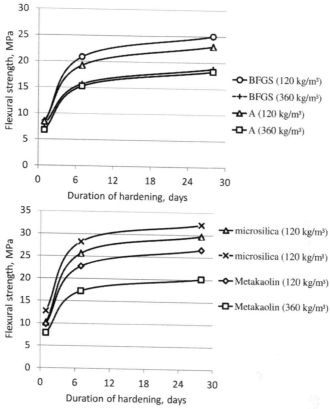

Figure 5.19. Kinetics of RPC Strength Growth for Different Types of Active Mineral Admixtures.

Table 5.10. Effect of Metakaolin and Other Mineral Admixtures on Reactive Powder Concrete Strength Characteristics.

No.	Type of mineral admixture, kg/m³	W/C	Flaw, cm	$f_{c.tf}^{1}$, MPa	f_{cm}^{1}, MPa	$f_{c.tf}^{7}$, MPa	f_{cm}^{7}, MPa	$f_{c.tf}^{28}$, MPa	f_{cm}^{28}, MPa
						Concrete strength at different age			
Metakaolin content 5% by cement weight									
1	Ash (A)	0.29	30	7.21	21.0	17.09	75.8	22.77	115.6
2	Blast furnace granulated slag (BFGS)	0.29	30	6.32	24.4	16.82	75.5	22.26	114,0
Metakaolin content 10% by cement weight									
3	Ash (A)	0.29	30	7.65	29.2	18.87	79.9	25.28	126.2
4	Blast furnace granulated slag (BFGS)	0.29	30	7.57	24.7	17.53	78.9	19.76	125.0
Metakaolin content 15% by cement weight									
5	Ash (A)	0.29	30	7.12	20.1	16.91	73.1	21.81	111.0
6	Blast furnace granulated slag (BFGS))	0.29	30	6.68	19.1	16.02	72.2	20.90	107.8

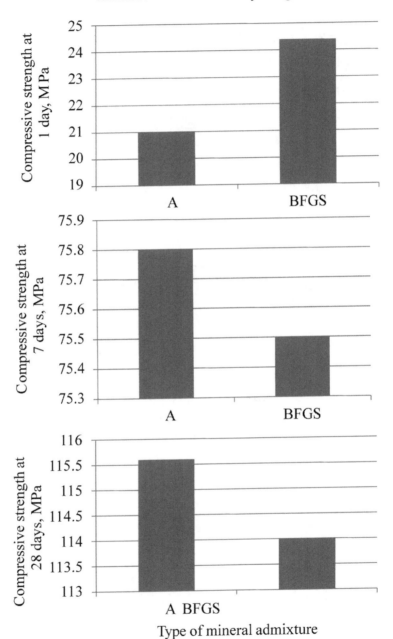

Figure 5.20. The Effect of 5% Metakaolin in Combination with other Mineral Admixtures on RPC Compressive Strength.

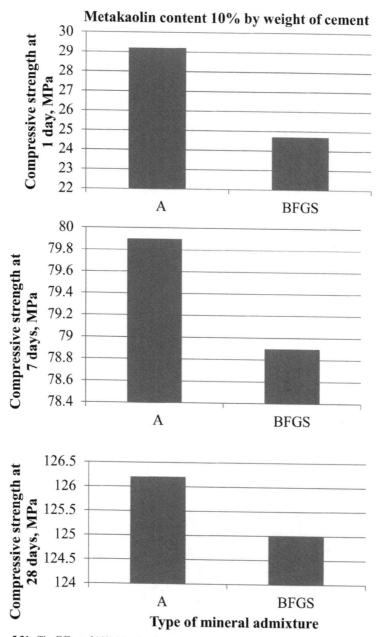

Figure 5.21. The Effect of 10% Metakaolin in Combination with other Mineral Admixtures on RPC Compressive Strength.

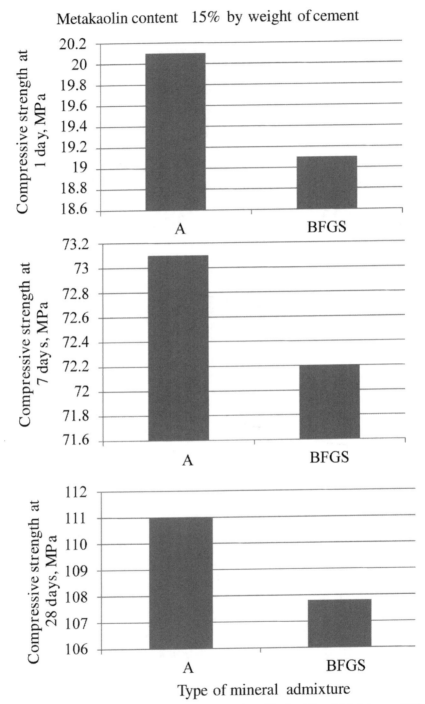

Figure 5.22. The Effect of 15% Metakaolin in Combination with other Mineral Admixtures on RPC Compressive Strength.

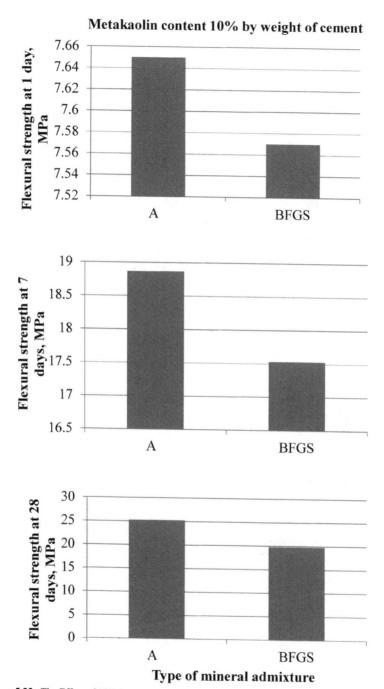

Figure 5.23. The Effect of 10% Metakaolin in Combination with other Mineral Admixtures on RPC Flexural Strength.

5.5 Optimal Compositions of Reactive Powder Concretes Containing Ash and Metakaolin

At this research stage the complex effect of cement consumption, content of active mineral admixtures and superplasticizer admixtures on reactive powder concrete strength was studied. The main studies were performed using mathematical experimental planning [13]. For this purpose, a three-level, three-factor B_3 plan was realized. Planning conditions and results of experimental studies are given in Tables 5.11, 5.12 and 5.13.

Portland cement CEM I 42.5N manufactured by the cement plant, in Ukraine, was used for the experimental research. The cement consumption was 720, 840 and 960 kg/m³ of concrete mix. The content of active mineral admixtures varied from 20 to 40% by cement weight. Fly ash of Burshtyn TPP and metakaolin (Western Kaolin Company Ltd.) were used as the active mineral admixtures.

Previous studies have shown that using these admixtures has a positive effect on reactive powder concrete strength, and their cost compared to microsilica as the main component of such concrete is significantly lower. The of influence composition of these admixtures on RPC strength is stronger, compared to other admixtures due to their synergistic effect.

Quartz sand fraction 0.16–1.25 mm was used as aggregate for RPC production. To minimize the water demand and increase the RPC strength, a polyacrylate superplasticizer type Dynamon SP-3 was added to the mixtures at mixing. The superplasticizer content varied from 1 to 2% by binder weight in accordance with the experimental planning conditions. Water demand at each point of the plan was determined from the condition of providing a given flow of 25…30 cm on the Suttard viscometer.

After experimental data processing and statistical analysis (Tables 5.12, 5.13), mathematical models of water demand, water-cement ratio, RPC compressive and flexural strength at 1, 7, and 28 days were obtained in a form of polynomial regression equations (Tables 5.14, 5.15).

Table 5.11. Experimental Planning Conditions.

No.	Factors		Variation levels			interval
	Coded	Natural	−1	0	+1	
1	X_1	Cement consumption (C), kg/m³	720	840	960	120
2	X_2	Mineral admixture to cement ratio (ma/C) by weight	0.2	0.3	0.4	0.1
3	X_3	Consumption of superplasticizer (SP) Dynamon SP-3, %	1	1.5	2	0.5

Table 5.12. Experimental Planning Matrix and Concrete Consumptions.

No.	X_1	X_2	X_3	C, kg/m³	Ash or metakaolin kg/m³	Sand, kg/m³	SP, kg/m³
1	1	1	1	960	384	926	26.88
2	1	1	−1	960	384	898	26.88
3	1	−1	1	960	192	1138	23.04
4	1	-1	−1	960	192	1118	23.04
5	−1	1	1	720	288	1303	20.16
6	−1	1	−1	720	288	1289	20.16
7	−1	−1	1	720	144	1469	17.28
8	−1	−1	−1	720	144	1447	17.28
9	1	0	0	960	288	1013	24.96
10	−1	0	0	720	216	1375	18.72
11	0	1	0	840	336	1100	23.52
12	0	−1	0	840	168	1285	20.16
13	0	0	1	840	252	1210	21.84
14	0	0	−1	840	252	1193	21.84
15	0	0	0	840	252	1193	21.84
16	0	0	0	840	252	1193	21.84
17	0	0	0	840	252	1193	21.84

Table 5.13. Results of Experimental Studies of the Concrete Composition Effect on its Strength (Mineral Admixture-fly Ash).

No.	W, l/m³	W/B**	W/C	$f_{c.tf}^1$, MPa	f_{cm}^1, MPa	$f_{c.tf}^7$, MPa	f_{cm}^7, MPa	$f_{c.tf}^{28}$, MPa	f_{cm}^{28}, MPa
1	250	0.19	0.26	8.2	32.9	13.7	72.3	23.6	111.8
2	278	0.21	0.29	7.7	29.7	12.4	57.6	22.5	99.3
3	230	0.20	0.24	8.9	35.6	14.9	73.1	23.3	112.1
4	250	0.22	0.26	7.8	32.2	13.8	62.1	23.8	101.3
5	209	0.21	0.29	7.0	25.4	12.8	56.6	22.5	104.6
6	223	0.22	0.31	6.0	21.9	11.5	55.8	21.4	93.6
7	187	0.22	0.26	7.6	28.1	14.1	57.3	22.2	105.5
8	209	0.24	0.29	6.2	24.4	12.9	60.4	22.8	96.2
9	259	0.21	0.27	8.7	36.8	15.1	70.4	24.1	115.2
10	209	0.22	0.29	7.2	29.2	14.2	61.7	23.0	108.4
11	244	0.21	0.29	7.4	30.9	12.6	61.1	22.3	106.7
12	227	0,23	0.27	7.7	33.5	13.9	63.8	22.8	108.2
13	218	0.20	0.26	8.2	35.7	14.6	65,5	23.2	114.7
14	235	0.22	0.28	7.2	32.3	13.4	59.6	22.8	103.8
15	235	0.22	0.28	7.9	35.2	14.3	64.6	23.1	112.6
16	235	0.22	0.28	7.9	35.2	14.3	64.5	23.2	112.6
17	235	0.22	0.28	7.9	35.2	14.4	64.6	23.2	112.6

* - concrete composition is given in Table 5.5
** W/B – water-binder ratio

Table 5.14. Results of Experimental Studies on the Concrete Composition Effect on its Strength (Mineral Admixture – Metakaolin).

No	W, l/m^3	W/B	W/C	$f_{c.tf}^1$, MPa	f_{cm}^1, MPa	$f_{c.tf}^7$, MPa	f_{cm}^7, MPa	$f_{c.tf}^{28}$, MPa	f_{cm}^{28}, MPa
1	251	0.19	0.26	8.9	35.5	19.7	90.1	23.2	117.2
2	303	0.23	0.32	7.2	32.1	15.9	72.5	18.6	94.3
3	236	0.21	0.25	9.4	41.2	20.7	94.4	24.3	122.8
4	248	0.22	0.26	8.5	33.8	18.8	86.1	22.2	112.0
5	209	0.21	0.29	8.3	32.9	18.4	84.1	21.6	109.2
6	247	0.24	0.34	6.8	30.1	14.8	67.6	17.4	88.0
7	200	0.23	0.28	8.8	38.8	19.5	88.9	22.9	115.5
8	198	0.23	0.27	8.1	32.0	17.9	81.7	21.0	106.3
9	263	0.21	0.27	9.6	40.1	21.2	96.6	24.9	125.6
10	217	0.23	0.30	9.0	37.9	20.0	91.4	23.5	118.8
11	260	0.22	0.31	8.0	34.3	18.1	82.5	21.2	107.3
12	228	0.23	0.27	8.9	38.1	20.1	91.7	23.6	119.3
13	220	0.20	0.26	9.5	40.7	21.5	98.1	25.2	127.4
14	245	0.22	0.29	8.3	35.5	18.8	85.7	22.0	111.4
15	240	0.22	0.29	9.2	39.4	20.8	94.9	24.4	123.3
16	240	0.22	0.29	9.2	39.4	20.8	94.7	24.4	123.3
17	240	0.22	0.29	9.2	39.4	20.8	95.0	24.4	123.3

* - concrete composition is given in Table 5.12
** W/B – water-binder ratio

Graphic dependences that illustrate the influence of technological factors on the water–cement ratio and concrete compressive and flexural strengths at 1 and 28 days of normal hardening are shown in Figs. 5.24–5.30.

It was shown that reducing the water-cement ratio of RPC produced using both fly ash and metakaolin leads to an increase in cement consumption and plasticizer content. With an increase in the content of these admixtures in RPC composition, the average decrease in W/C is 10… 12%. Some increase in W/C is associated with a decrease in the total amount of cement in the binder. It should also be noted that when using the fly ash as active mineral admixture for RPC there is a decrease in W/C at all plan points by an average of 5–7% compared with metakaolin. This is due to the spherical shape of ash particles, as well as their reduced porosity compared to slag, which accordingly leads to an additional plasticizing effect. Metakaolin, on the contrary, due to its dispersed clay layered particles causes a much higher increase in concrete water demand. Simultaneous increase of the superplasticizer content (interaction coefficient at X_2X_3) causes some compensation of the water demand increase.

Table 5.15. Experimental Statistical Models of Water Demand, Water-cement Ratio and RPC Strength.

RPC produced using TPP fly ash	
Water demand, l/m^3	$W = 234.7 + 23X_1 + 10.1X_2 - 10.1X_3 + 0.2X_1^2 + 1.7X_2^2 - 7.3X_3^2 + 1.5X_1X_2 - 1.5X_1X_3$ $\hfill(5.41)$
Water – cement ratio	$W/C = 0.28 - 0.012X_1 + 0.012X_2 - 0.012X_3 + 0.002X_1^2 + 0.002X_2^2 - 0.008X_3^2$ $\hfill(5.42)$
Flexural strength at 1 day, MPa	$f_{c.tf}^{\ 1} = 7.94 + 0.722X_1 - 0.19X_2 + 0.5X_3 + 0.065X_1^2 - 0.318X_2^2 - 0.205X_3^2 - 0.1X_1X_2 - 0.113X_2X_3$ $\hfill(5.43)$
Compressive strength at 1 day, MPa	$f_{cm}^{\ 1} = 35.28 + 3.82X_1 - 1.3X_2 + 1.72X_3 - 2.2X_1^2 - 3.03X_2^2 - 1.23X_3^2 - 0.075X_1X_3 - 0.05X_2X_3$ $\hfill(5.44)$
Flexural strength at 7 days, MPa	$f_{c.tf}^{\ 7} = 14.38 + 0.455X_1 - 0.66X_2 + 0.611X_3 + 0.292X_1^2 - 1.06X_2^2 - 0.338X_3^2 + 0.01X_1X_2 - 0.015X_1X_3 + 0.045X_2X$ $\hfill(5.45)$
Compressive strength at 7 days, MPa	$f_{cm}^{\ 7} = 64.55 + 4.38X_1 - 1.32X_2 + 2.92X_3 + 1.48X_1^2 - 2.12X_2^2 - 2X_3^2 - 3.5X_1X_3 + 0.95X_2X_3$ $\hfill(5.46)$
Flexural strength at 28 days, MPa	$f_{c.tf}^{\ 28} = 23.21 + 0.55X_1 - 0.25X_2 + 0.14X_3 + 0.4X_1^2 - 0.6X_2^2 - 0.13X_3^2 + 0.4X_2X_3$ $\hfill(5.47)$
Compressive strength at 28 days, MPa	$f_{cm}^{\ 28} = 112.9 + 3.1X_1 - 0.7X_2 + 5.5X_3 - 1X_1^2 - 5.3X_2^2 - 3.5X_3^2 + 0.2X_1X_2 + 0.4X_1X_3 + 0.4X_2X_3$ $\hfill(5.48)$
RPC produced using metakaolin	
Water demand, l/m^3	$W = 240 + 23.0X_1 + 15.9X_2 - 12.4X_3 + 0.19X_1^2 + 3.7X_2^2 - 7.3X_3^2 + 1.5X_1X_2 - 3.6X_1X_3 - 9.8X_2X_3$ $\hfill(5.49)$
Water – cement ratio	$W/C = 0.29 - 0.012X_1 + 0.019X_2 - 0.014X_3 + 0.005X_1^2 - 0.01X_3^2 - 0.014X_2X_3$ $\hfill(5.50)$
Flexural strength at 7 days, MPa	$f_{c.tf}^{\ 1} = 9.2 + 0.26X_1 - 0.45X_2 + 0.6X_3 + 0.09X_1^2 - 0.75X_2^2 - 0.29X_3^2 + 0.03X_1X_3 + 0.2X_2X_3$ $\hfill(5.51)$
Compressive strength at 7 days, MPa	$f_{cm}^{\ 1} = 39.4 + 1.1X_1 - 1.9X_2 + 2.56X_3 - 0.37X_1^2 - 3.2X_2^2 - 1.3X_3^2 + 0.05X_1X_2 + 0.13X_1X_3 - X_2X_3$ $\hfill(5.52)$
Flexural strength at 28 days, MPa	$f_{c.tf}^{\ 7} = 20.8 + 0.57X_1 - X_2 + 1.35X_3 - 0.19X_1^2 - 1.7X_2^2 - 0.7X_3^2 + 0.03X_1X_2 + 0.07X_1X_3 + 0.5X_2X_3$ $\hfill(5.53)$
Compressive strength at 28 days, MPa	$f_{cm}^{\ 7} = 94.9 + 2.6X_1 - 4.6X_2 + 6.2X_3 - 0.89X_1^2 - 7.8X_2^2 - 3.03X_3^2 + 0.13X_1X_2 + 0.3X_1X_3 + 2.31X_2X_3$ $\hfill(5.54)$
Flexural strength at 7 days, MPa	$f_{c.tf}^{\ 28} = 24.4 + 0.68X_1 - 1.19X_2 + 1.6X_3 - 0.23X_1^2 - 2X_2^2 - 0.78X_3^2 + 0.03X_1X_2 + 0.08X_1X_3 + 0.6X_2X_3$ $\hfill(5.55)$
Compressive strength at 7 days, MPa	$f_{cm}^{\ 28} = 123.3 + 3.4X_1 - 6.1X_2 + 8.2X_3 - 1.16X_1^2 - 10.1X_2^2 - 3.93X_3^2 + 0.16X_1X_2 + 0.41X_1X_3 + 5.0X_2X_3$ $\hfill(5.56)$

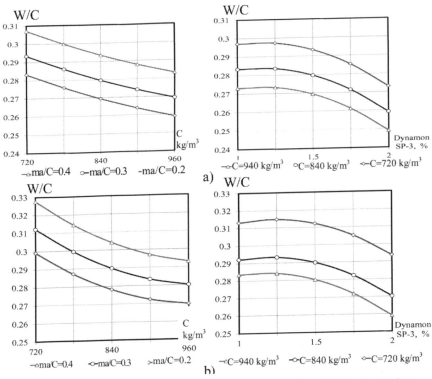

Figure 5.24. Effect of Technological Factors on Water-cement ratio of RPC Produced using:
a) TPP fly ash; b) Metakaolin.

As follows from the analysis of experimental and statistical models as well as corresponding graphical dependences, the nature of the varied factors effect on RPC strength characteristics, does not change significantly with increasing the concrete hardening duration. The increase in the compressive and flexural strength of concrete is due to higher cement consumption and addition of Dynamon SP-3 superplasticizer, which is mainly due to a sharp decrease in water-cement ratio and a corresponding increase in the specimens' density. The influence of factor X_2 (the degree of binder filling with mineral admixture) on the RPC strength characteristics has an extreme nature. In other words, an increase in the amount of active mineral admixture from 20 to 30% by binder weight increases the strength. A further increase in its amount causes a sharp decrease in strength, which is explained by a decrease in the amount of active clinker component in the total weight of binder. A similar effect is observed when using ash and metakaolin.

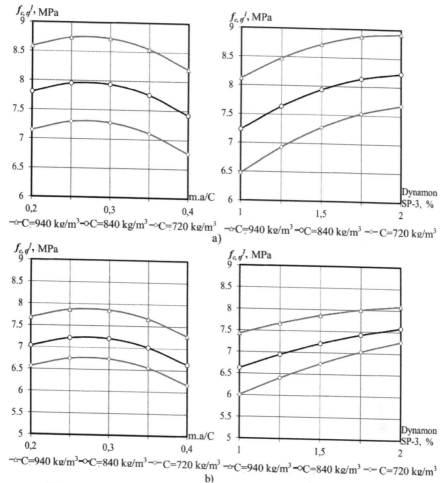

Figure 5.25. Effect of Technological Factors on Flexural Strength of RPC at 1 Day:
a) RPC produced using of TPP Fly Ash; b) RPC produced using Metakaolin.

Analysing the obtained experimental data it was found that metakaolin is a more effective active mineral admixture(from the two investigated) for producing RPC. Replacing metakaolin by the same amount of fly ash leads an average decrease in the strength of specimens by 12 – 15%. The maximum RPC compressive and flexural strengths of 131.2 MPa and 25.2 MPa were observed at the following ratio of variable factors: cement consumption at the maximum level – 960 kg/m³, the ratio between the mineral admixture content (fly ash)

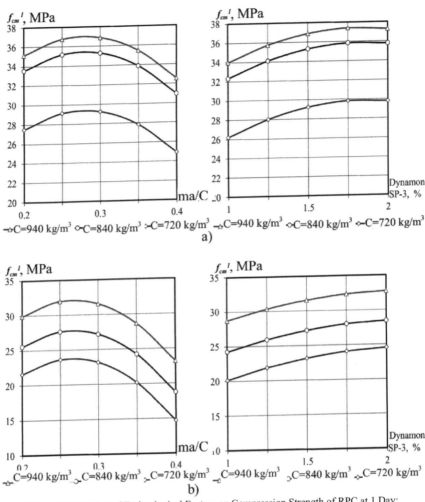

Figure 5.26. Effect of Technological Factors on Compression Strength of RPC at 1 Day:
a) RPC produced using of TPP Fly Ash; b) RPC Produced using Metakaolin.

and cement consumption at average levels – 0.3; content of Dynamon SP-3 superplasticizer at the maximum level – 2% by binder weight.

Based on the experimental-statistical models (5.41, 5.50, 5.51, 5.56) (Table 5.16), nomograms of water demand and strength of RPC produced using TPP fly ash and metakaolin as an active mineral admixture were obtained. These nomograms together with a set of obtained models (Table 5.16), can be used to design RPC compositions with a given strength and fluidity. The design methodology includes:

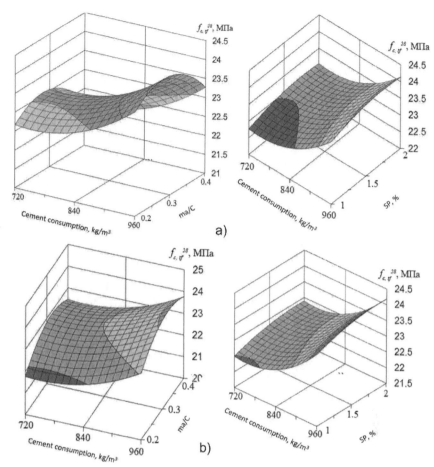

Figure 5.27. Effect of Technological Factors on Flexural Strength of RPC at 28 Days:
a) RPC Produced using of TPP Fly Ash; b) RPC Produced using Metakaolin.

1. Using the nomogram of compressive strength shown in Fig. 5.29 or 5.30, depending on the type of the admixture, determine the content of superplasticizer Dynamon SP-3, cement consumption and admixture content, which provide the specified RPC strength at 28 days.

2. Using the water demand nomogram shown in Figs. 5.29 or 5.30, depending on the type of mineral admixture used, at predetermined consumptions of superplasticizer, cement and admixture, set the water demand that will provide a concrete mixture with a fluidity of 25 – 30 cm by a Suttard viscometer.

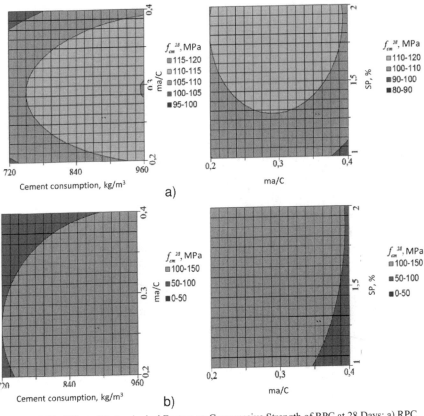

Figure 5.28. Effect of Technological Factors on Compressive Strength of RPC at 28 Days: a) RPC Produced using of Ash; b) RPC Produced using Metakaolin.

3. The aggregates consumption can be calculated knowing the cement paste volume ($V_{c.p}$) in the concrete mix. The cement paste volume, l/m^3 is

$$V_{c.p} = \frac{C}{\rho_c} + \frac{m.a}{\rho_{m.a}} + W \qquad (5.57)$$

The volume of sand, l/m^3 is

$$V_S = 1000 - V_{c.p.} \qquad (5.58)$$

The sand weight S, kg/m^3 is

$$S = \rho_s V_s \qquad (5.59)$$

In the above equations ρ_c, $\rho_{m.a}$, ρ_s are the real densities of cement ($\rho_c \approx 3.1$ kg/l), mineral admixture and sand, respectively.

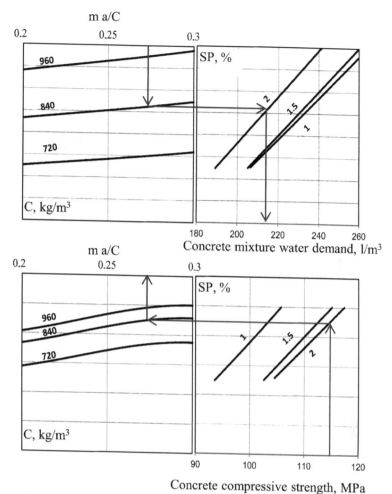

Figure 5.29. Nomograms of Water Demand and Compressive Strength of RPC made Using Fly Ash.

Calculation Example

Calculate the composition of RPC, produced using TPP fly ash as active mineral admixture. The target compressive strength at 28–days is 115 MPa, the required concrete mix fluidity by Suttard viscometer is 25 – 30 cm. As the plasticizing admixture is used Dynamon SP-3 superplasticizer. The real densities are as follows:

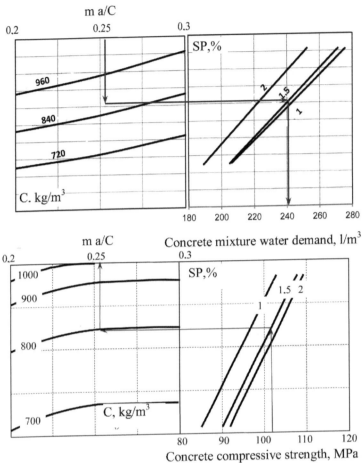

Figure 5.30. Nomogram of Water Demand and Compressive Strength of RPC Produced using Metakaolin.

- cement $\rho_c = 3.1$ g/cm³,
- ash $\rho_{m.a} = 2.8$ g/cm³,
- sand $\rho_s = 2.65$ g/cm³.

1. Following the nomogram for compressive strength shown in Fig. 5.29, to achieve the specified compressive strength of 110 MPa, the minimum possible cement consumption is 840 kg/m³, admixture content – 27.5% by cement weight, content of Dynamon SP-3 superplasticizer – 2% by binder weight.

2. Using the nomogram for water demand shown in Fig. 5.29, for the calculated concrete composition of minimum possible water demand providing the required concrete mix fluidity of 25 – 30 cm by Suttard viscometer is 215 l/m³.

3. The aggregates consumption is calculated for the known cement paste volume ($V_{c.p}$) in the concrete mixture.

The cement paste volume, l/m^3 is

$$V_{c.p} = \frac{C}{\rho_c} + \frac{m.a}{\rho_{m.d}} + W = \frac{840}{3,1} + \frac{231}{2,8} + 215 = 568.5 \; l/m^3$$

The volume of sand, l/m^3 is

$$V_s = 1000 - V_{c.p} = 1000 - 568.5 = 431.5$$

The sand weight S, kg/m^3 is

$$S = \rho_c V_s = 2.65 \cdot 431.5 = 1143 \; kg/m^3$$

The concrete mix composition is: cement – 840 kg/m^3, fly ash – 231 kg/m^3, water – 215 l/m^3, sand fraction 0.16... 1.25 – 1143 kg/m^3. The content of Dynamon SP-3 superplasticizer is 2% by binder weight.

As noted in subsection 5.1, when using fly ash as an active mineral admixture for RPC adding metakaolin gives an increase in its action. To assess the complex effect of fly ash and metakaolin on the mechanical properties of RPC, a series of experiments was carried out according to the mathematical plan B_4 [13]. The effect of the mineral admixture (m.a), which was a mechanical mixture of fly ash and metakaolin, was investigated. Factor X_4 characterized the qualitative composition of this admixture—the ash and metakaolin mass ratio. As other variable factors in the experiment, the following were selected: cement consumption (X_1), the admixture and cement ratio (A/C (X_2)) and the consumption of Dynamon SP-3 superplasticizer (X_3). The levels of factors change and their designations are given in Table 5.16. The experimental results are shown in Table 5.17.

Table 5.16. Experimental Planning Conditions.

No.	Factors		Variation levels			Interval
	Coded	Natural	−1	0	+1	
1	X_1	Cement consumption (C), kg/m^3	720	840	960	120
2	X_2	Admixture to cement ratio (ma/C) by mass	0.2	0.3	0.4	0.1
3	X_3	Consumption of superplasticizer (SP) Dynamon SP-3, %	1	1.5	2	0.5
4.	X_4	Ash and metakaolin mass ratio in admixture (A/M)	0	0.5	1	0.5

Table 5.17. Experiment Planning Matrix, RPC ComponentConsumptions and Specimens Strength.

No.	X_1	X_2	X_3	X_4	C, kg/m³	Ash, kg/m³	Metakaolin, kg/m³	Sand, kg/m³	SP, kg/m³	Compressive strength, MPa	
										1 day	28 days
1	1	1	1	1	960	384	0	945	26.9	24.4	85.2
2	1	1	1	-1	960	0	384	945	26.9	42.8	124.2
3	1	1	-1	1	960	384	0	945	26.9	19.5	73.2
4	1	1	-1	-1	960	0	384	945	26.9	33.4	88.2
5	1	-1	1	1	960	192	0	1137	23.0	26.0	86.5
6	1	-1	1	-1	960	0	192	1137	23.0	38.0	109.5
7	1	-1	-1	1	960	192	0	1137	23.0	26.7	95.4
8	1	-1	-1	-1	960	0	192	1137	23.0	34.2	94.4
9	-1	1	1	1	720	288	0	1281	20.2	21.7	76.6
10	-1	1	1	-1	720	0	288	1281	20.2	38.2	109.6
11	-1	1	-1	1	720	288	0	1281	20.2	18.3	69.5
12	-1	1	-1	-1	720	0	288	1281	20.2	30.4	78.5
13	-1	-1	1	1	720	144	0	1425	17.3	24.7	82.0
14	-1	-1	1	-1	720	0	144	1425	17.3	34.7	99.0
15	-1	-1	-1	1	720	144	0	1425	17.3	26.9	95.8
16	-1	-1	-1	-1	720	0	144	1425	17.3	32.5	88.8
17	1	0	0	0	960	96	96	1137	23.0	37.4	117.1
18	-1	0	0	0	720	144	144	1281	20.2	35.2	110.0
19	0	1	0	0	840	126	126	1197	21.8	34.6	106.2
20	0	-1	0	0	840	126	126	1197	21.8	36.5	112.0
21	0	0	1	0	840	126	126	1197	21.8	38.0	118.7
22	0	0	-1	0	840	126	126	1197	21.8	34.4	107.7
23	0	0	0	1	840	252	0	1197	21.8	25.7	91.3
24	0	0	0	-1	840	0	252	1197	21.8	37.7	107.3

As a result of the statistical processing of the data (Table 5.18), regression equations for the strength of RPC in compression at the age of 1 day and 28 days were obtained:

$$f_{cm}^{1} = 36.7 + 1.1X_1 - 0.93X_2 + 1.79X_3 - 5.91X_4 - 0.44X_1^2 - 1.2X_2^2 - 0.55X_3^2$$
$$- 4.9X_4^2 + 0.34X_1X_2 +$$
$$+ 0.39X_1X_3 - 0.48X_1X_4 + 1.4X_2X_3 - 1.6X_2X_4 - 1.1X_3X_4 \qquad (5.60)$$

$$f_{cm}^{28} = 114.7 + 3.6X_1 - 2.9X_2 + 5.6X_3 - 8.06X_4 - 1.38X_1^2 - 5.8X_2^2 - 1.72X_3^2$$
$$- 1.56X_4^2 + 1.05X_1X_2 +$$
$$+ 1.21X_1X_3 - 1.5X_1X_4 + 5.2X_2X_3 - 4.1X_2X_4 - 6.1X_3X_4 \qquad (5.61)$$

Analyzing the equations obtained, it can be argued that the compressive strength of RPC with ashmetakaolin admixture for 1 day was in the range from 23 to 39 MPa, and on 28 – from 78 to 121 MPa. These results show that this complex mineral admixture allows obtaining RPC with a strength that is close to the strength values of RPC from silica fume (subsection 5.1).

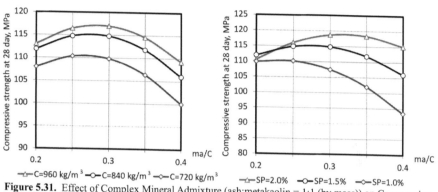

Figure 5.31. Effect of Complex Mineral Admixture (ash:metakaolin = 1:1 (by mass)) on Compressive Strength of RPC at 28 Days.

Among the factors investigated, the greatest individual effect on strength at 28 days is exerted by factor X_4 (the qualitative composition of the investigated complex admixture), other factors have a comparatively less effect (Fig. 5.31). Among the factors X_1 ... X_3, X_2 (the proportion of the mineral admixture) and cement consumption are more noticeable. The factor of fly ash ratio in a mixture with metakaolin (X_4), which characterizes the qualitative composition of the active mineral admixture, has a pronounced extreme effect on strength, that is, it indicates the existence of an optimal metakaolin and fly ash ratio in the composition of the complex admixture, and causes some synergistic effect. As can be seen from the constructed graphical dependencies, in the case when this mineral admixture is represented to a greater extent by fly ash, the compressive strength is within 78 ... 92 MPa. The maximum amount of metakaolin causes an increase in the strength of the RPC to 93 ... 117 MPa. The maximum strength is achieved with a ratio of ash to metakaolin from 23 to 57%. Fluctuations in this ratio are caused by a change in other factors: to the greatest extent by the consumption of the superplasticizer (X_3) and the proportion of the active mineral admixture in the mixture with cement (X_2), to a lesser extent—by the consumption of cement (X_1).

Graphs of changes in the optimal ash and metakaolin ratio from the point of view of achieving maximum strength at 28 days of hardening on recall surfaces are shown in Fig. 5.32 and 5.33. Also, the coefficients of the equations show that the factor characterizing the content of active mineral admixture in the mixture (X_3) strongly interacts with the consumption of the superplasticizer. Such powerful interactions of these factors indicate that the main reason for the effect of complex ash metakaolin admixtures on strength is its effect on the rheological

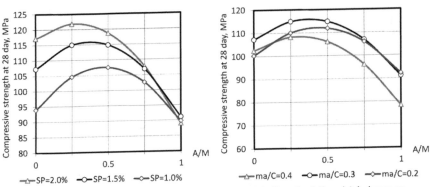

Figure 5.32. Effect of Ash and Metakaolin Ratio (A/M) in Complex Mineral Admixture on Compressive Strength of RPC at 28 Days.

properties of RPC. As noted earlier [25] metakaolin, due to its high dispersion and structural structure, has a sufficiently high water demand, and therefore, in order to ensure a dense structure of cement stone filled with metakaolin, it is necessary to compensate for the increase in viscosity that occurs in this case. Such compensation is carried out by increasing the content of the superplasticizer, which is reflected through an increase in the strength of the RPC. Fly ash has a significantly lower pozzolanic activity than metakaolin, however, due to the spherical shape of the watch [2, 5, 7], it reduces the viscosity of the cement paste, especially in conjunction with a superplasticizer. Thus, as a result of the action of the triple complex metakaolin-ash-superplasticizer, in which the superplasticizer and fly ash neutralize the effect of metakaolin on the rheological properties of the concrete mixture, the maximum values of the strength of the RPB are achieved. At the maximum content of Dynamon SP-3 superplasticizer (2% of the cement

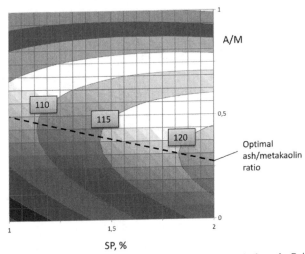

Figure 5.33. Graphs of Changes in the Optimal Ash and Metakaolin Ratio from the Point of Achieving Maximum Compressive Strength at 28 Days (MPa) View.

mass), the maximum strength (121.7 MPa) is shifted towards a greater amount of Metakaolin in the composition; at the minimum level of the superplasticizer (1%), the highest strength of 107 MPa is achieved with an approximately equal ratio of active mineral components.

On the first day of RPC hardening (Fig. 5.34) with ash-metakaolin mineral admixture, the nature of the influence of factors varying in the experiment is the same as on the 28th. Significant shift of the optimal from the standpoint of maximum strength ratio between the active components in the direction of a higher content of metakaolin, as the reaction of the fly ash with calcium hydroxide is much slower.

5.6 Influence of Hardening Temperature Modes on Reactive Powder Concrete Strength

Research results have shown that RPC properties are significantly affected by using additional pressure at molding, changes in curing temperature, as well as joint use of pressure and temperature. Using pressure leads to reduction of entrained air, removal of excessing water and chemical shrinkage compensation [198]. Similar phenomena were observed at compression when the investigated specimens of the RPC were subjected to a given load. It was shown that the air and water voids present in RPC without additional pressure, are significantly reduced when a given pressure is applied. As known [199], significantly increase in concrete compressive strength and is obtained by casting concrete in steel tubes. This method is also effective for RPC [200]. It is found that RPC hardening in confined conditions due to the use of steel tubes significantly increases its compressive strength, as well as longitudinal deformation. A similar result is obtained for concrete impact strength that significantly improves when steel tube confinement is used [198].

Heat processing of RPC after molding is quite effective [201] in terms of accelerating the pozzolanic reaction of microsilica and quartz. It significantly changes the RPC microstructure and also has a positive effect on the process of excessive water removing. Heat processing after hardening promotes formation of tobermorite ($Ca_5Si_6O_{16}(OH)_2 \times 4H_2O$) at temperatures up to 200°C, and xonolite ($Ca_6Si_6O_{17}(OH)_2$ at temperatures of 250°C and higher). The presence of xonolite in RPC at 250°C was confirmed in research [198]. It was observed that increasing the temperature accelerates the hydration process at an early age and increases the compressive strength of the material. However, the rapid hydration process leads to a fast increase in autogenous shrinkage of concrete at early age.

A series of studies was performed to determine the effect of heat processing on the curing process and the strength of RPC. For all experiments that have been performed, the RPC composition was as follows: cement consumption – 840 kg/m³, mineral admixtures – 360 kg/m³, sand fraction 0.16 – 1.25 mm – 1200 kg/m³, polyacrylate type Dynamon SP -3 superplasticizer – 2% by binder weight. Water demand was obtained from the condition of providing a given fluidity of

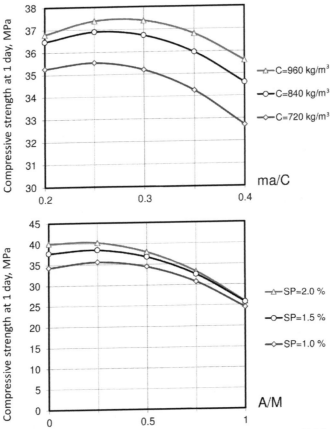

Figure 5.34. Effect of Complex Mineral Admixture on Compressive Strength of RPC at 1 Day.

25 – 30 cm by Suttard viscometer. As a mineral admixture at this research stage was used fly ash (FA). Additionally, metakaolin was added to the concrete mixture (10% of cement weight). Heat processing was carried out in the laboratory steaming chamber according to the following mode: preliminary exposure – 2 h; temperature rise at a rate of 25°C/h; isothermal exposure – 8 h, cooling. After heat processing, the compressive strength of the specimens was determined at 2 h, 7 days and 28 days. The strength values are given in Table 5.18.

In some research, along with steaming, to ensure maximum RPC hydration and pozzolanic reaction, hardening in hot water is used. To determine the RPC curing efficiency the investigated specimens were kept in water at a temperature of 50 and 80°C.

The strength of RPC specimens hardened in hot water are given in Table 5.19. RPC steaming at a temperature of 50°C causes an increase in strength at 1 day by 2.5 – 3 times, at 28 days – by 1.5 – 2%. Increasing the isothermal heating temperature causes an increase in the RPC strength. Compared with specimens that hardened under normal conditions, steaming at 80°C causes an increase in

Table 5.18. Influence of Heat and Humid Processing on the RPC Strength Characteristics.

No.	Type of mineral admixture, kg/m³	W/C	Flaw, cm	Concrete strength					
				$f_{c.tf}^{HP}$, MPa	f_{cm}^{HP} MPa	$f_{c.tf}^{7}$, MPa	f_{cm}^{7}, MPa	$f_{c.tf}^{28}$, MPa	f_{cm}^{28}, MPa
	Isothermal exposure temperature 50°C								
1	Fly ash (FA)	0.29	30	14.64	51.2	22.45	93.7	24.14	115.6
2	Flyash (FA)+metakaolin (MTK)	0.29	30	17.38	72.4	24.40	99.2	26.47	127.6
	Isothermal exposure temperature 80°C								
3	Fly ash (FA)	0.29	30	19.90	79.7	24.07	96.1	25.39	120.0
4	Flyash (FA)+metakaolin (MK)	0.29	30	26.15	92.6	27.23	105.3	28.48	132.6

Table 5.19. Influence of Heat and Humid Processing by Hot Water on RPC Strength Characteristics.

No.	Type of mineral admixture, kg/m3	W/C	Flow, cm	Concrete strength					
				$f_{c.tf}^{HP}$, MPa	f_{cm}^{HP} MPa	$f_{c.tf}^{7}$, MPa	f_{cm}^{7}, MPa	$f_{c.tf}^{28}$, MPa	f_{cm}^{28}, MPa
	Isothermal exposure temperature 50°C								
1	Ash (A)	0.29	30	15.44	52.2	23.48	98.4	25.17	121.7
2	Ash (A)+metakaolin (MTK)	0.29	30	18.42	73.6	25.85	106.2	27.66	135.8
	Isothermal exposure temperature 80°C								
3	Ash (A)	0.29	30	21.64	91.0	26.92	103.9	27.21	134.4
4	Ash (A)+metakaolin (MTK)	0.29	30	28.91	103.9	29.09	115.7	30.82	145.9

strength at 1 day – by 3.5 ... 4 times, at 28 days – by 4 – 5%. As known, heat and humid processing significantly activates the pozzolanic activity of various mineral admixtures. The obtained results show that steaming significantly increases the effect of strength growth caused by addition of fly ash. At 50 – 80°C heat processing just accelerates the interaction of calcium hydroxide with amorphous silica ash. The positive effect of steaming is also reflected in the flexural strength.

The maximum values of RPC strength were obtained for specimens that cured in water at a 80°C (Table 5.17). RPC, which contained a complex mineral admixture—10% of metakaolin and 10% of fly ash of cement weight, reached after 28 days of hardening compressive and flexural strength values of 145 MPa and 30.8 MPa, respectively, which is 15 – 17% higher than for concrete with the same composition that hardened under normal conditions.

The obtained results on the influence of different types of mineral additives (high-activity metakaolin and fly ash) and their compositions on compressive and flexural strength of RPC as well as kinetics of its strength growth prove the possibility of producing concrete with the compressive strength in the range of 100 – 133 MPa.

It was found that a combination of metakaolin with ash is an effective composite mineral additive, allowing replacing microsilica in RPC. The compressive strength of RPC with this composition under normal curing conditions is 124 – 126 MPa and after exposing in water at 80°C it reaches 145 MPa. The synergistic effect of this composite additive is explained by the positive effect of spherical ash particles on the concrete mixtures workability, which to some extent compensates for the increase in viscosity caused by the addition of metakaolin.

References

1. Bazhenov, Yu. M. 1987. Tekhnologiya betona [Concrete technology]. Moscow: Vysshaya shkola, 415 p. (In Russian).
2. Kokubu, M. 1976. Tsement s dobavleniyem letuchey zoly (osnovnoy doklad) [Cement with the addition of fly ash (main report)]. Sixth International Congress on Cement Chemistry: Proceedings. Moscow: Stroyizdat, pp. 405–416. (In Russian).
3. Sergeev, A.M. 1984. Ispol'zovaniye v stroitel'stve otkhodov energeticheskoy promyshlennosti. [Use in construction of waste of the power industry], Kyiv: Budivelnik. 120 p. (In Russian).
4. Dvorkin, L. and O. Dvorkin. 2006. Basic of concrete science: Optimum Design of Concrete Mixtures. Kindle Edition. 237 p.
5. Lyur, H.P. and Efas, Ya. 1976. Vpliv granulometrichnogo skladu zol z nizkimi vtratami pri prozharyuvanni na zrostannya mitsnosti betonu, tehnologiya tovarnoyi betonnoyi sumishi [The dependence of concrete strength from the granulometric composition of ash with low loss while heating], Sixth International Cement Chemistry Congress, Stroyizdat, Moscow, pp. 18–24. (In Russian).
6. Dvorkin, L.I., V.I. Solomatov, V.N. Vyrovoj and S.M. Chudnovskij. 1991. Czementnye betony s mineralnymi napolnitelyami [Cement-based concrete with mineral fillers]. Kyiv: Budivelnyk, 136 p. (In Russian).
7. Rekomendatsii po ispol'zovaniyu zoloshlakovykh i zoloshlakovykh smesey TES v betone [Recommendations for the use of ash, slag and ash-slag mixture of thermal power plants in concrete]. 1986. Moskow: NIIZHB. 78 p. (In Russian).
8. Stolnikov, V.V. 1989. Ispol'zovaniye letuchey zoly ot szhiganiya pylevidnogo topliva na teplovykh elektrostantsiyakh. [The use of fly ash from the combustion of pulverized fuel in thermal power plants]. Leningrad: Energy. 50 p. (In Russian).
9. Kovach, R. 1976. Protsessy gidratatsii i prochnosti zol'nykh tsementov. [Processes of hydration and durability of ash cements]//Sixth International Congress on Cement Chemistry: Proceedings. Vol.3. Moscow: Stroyizdat. pp. 91–102. (In Russian).
10. Kokubu, I. 1973. Zola i zol'nyye tsementy. Ash and ash cements. Fifth International Congress on Cement Chemistry. Moskow: Stroyizdat. pp. 405–416. (In Russian).
11. Venua, M. 1980. Tsement i beton v stroitel'stve. Cement and concrete in construction. Moscow: Stroyizdat. 415 p. (In Russian).
12. Gorchakov, G.I., A.B. Nabokov and S.F. Pritula. 1978. Struktura i morozostoykost' gidrotehnicheskogo betona s dobavleniyem zoly unosa teplovykh elektrostantsiy. [Structure and frost resistance of hydraulic concrete with the addition of fly ash from thermal power plants]. Proceedings of the conference. and meetings on hydraulic engineering. Issue. 118 Leningrad: Energy. pp. 80–85. (In Russian).
13. Dvorkin, L., O. Dvorkin and Y. Ribakov. 2013. Multi-Parametric Concrete Compositions Design. New York: Nova Science Pub Inc., 233 p.
14. Entin, Z.B. 1976. O gidratatsii i tverdenii tsementov s zoloy. [About hydration and hardening of cements with ash]. Sixth International Congress on Cement Chemistry: Proceedings. Vol. 3. Moscow: Stroyizdat. pp. 95–99. (In Russian).
15. Stolnikov, V.V. and R.E. Litvinova. 1972. Treshchinostojkost' betona. Moscow: Energiya. (In Russian), 113 p.
16. Neville, A.M. 1996. Properties of concrete. 4th edition, Wiley & Sons, New York, 844 p.

17. Jepson, W.B. 1984. Kaolins: Their Properties and Uses. Philosophical Transactions of the Royal Society of London. Series A, Mathematical and Physical Sciences 311(1517): 411–32. Accessed November 20, 2020. http://www.jstor.org/stable/37333.

18. Murray, H.H. 2006. Kaolin Applications. Chapter 5 Developments in Clay Science edited by Haydn H. Murray. Elsevier Science Vol. 2: 85–109. doi.org/10.1016/S1572-4352(06)02005-8.

19. Yaya, A., E.K. Tiburu, M.E. Vickers, J.K. Efavi, B. Onwona-Agyeman and K.M. Knowles. 2017. Characterisation and identification of local kaolin clay from Ghana: A potential material for electroporcelain insulator fabrication. Applied Clay Science 150: 125–130. doi.org/10.1016/j.clay.2017.09.015.

20. Clays. In: U.S. Geological Survey. 2020. Mineral commodity summaries 2020: U.S. Geological Survey. doi.org/10.3133/.

21. Kaolin (China clay). Accessed November 21, 2020. https://www.mindat.org/min-52467.html.

22. Brindley, G. and K. Robinson. 1946. The structure of kaolinite. Mineralogical Magazine and Journal of the Mineralogical Society 27(194): 242–253. doi: 10.1180/minmag.1946.027.194.04.

23. Ovcharenko, F., N. Kruhlytskiy, Yu. Rusko, I. Moroz, M. Komska and Yu. Teodorovych. 1982. Kaoliny Ukrainy: Spravochnik [Kaolins of Ukraine: Handbook]. Kyiv: Naukova dumka, 368 p. (in Russian)

24. Dvorkin, L., N. Lushnikova, R. Runova and V. Troyan. 2007. Metakaolin in building mortars and concrete [Metakaolin v budivelnych rozchynach i betonach]. Kyiv: KNUBA, 216 p. (In Ukrainian).

25. Kubliha, M., V. Trnovcová, J. Ondruška, I. Štubňa, O. Bošák and T. Kaljuvee. 2017. Comparison of dehydration in kaolin and illite using DC conductivity measurements. Applied Clay Science 149: 8–12.

26. Fernandez, R., F. Martirena and K.L. Scrivener. 2011. The origin of the pozzolanic activity of calcined clay minerals: A comparison between kaolinite, illite and montmorillonite. Cement and Concrete Research 41(1): 113–122.

27. Justice, Joy M. 2005. Evaluation of metakaolins for use as supplementary cementitious materials. PhD diss. Georgia Institute of Technology. https://smartech.gatech.edu/bitstream/handle/1853/6936/justice_joy_m_200505_mast.pdf?sequence=1&isAllowed=y.

28. Mastai, Yitzhak, ed. 2013. Material Science- Advanced Topics. In-Tech Open. doi: 10.5772/56700.

29. Tironi, A., M.A. Trezza, A.N. Scian and E.F. Irassar. 2014. Thermal analysis to assess pozzolanic activity of calcined kaolinitic clays. Journal of Thermal Analysis and Calorimetry 117: 547–556. doi.org/10.1007/s10973-014-3816-1.

30. Kaljuvee, T., I. Štubňa, T. Húlan and R. Kuusik. 2017. Heating rate effect on the thermal behavior of some clays and their blends with oil shale ash additives. Journal of Thermal Analysis and Calorimetry 127: 33–45. doi.org/10.1007/s10973-016-5347-4.

31. Percival, H.J., J. Duncan and P.K. Foster. 1974. Interpretation of the Kaolinite-Mullite Reaction Sequence from Infrared Absorption Spectra. Journal of the American Ceramic Society 57: 57–61.

32. Biljana, R., A. Aleksandra and R. Ljiljana. 2010. Thermal treatment of kaolin clay to obtain metakaolin. Hemijska Industrija 64: 351–356.

33. Khatib, J. M., O. Baalbaki and A.A. El Kordi. 2018. "15 – Metakaolin" in Woodhead Publishing Series in Civil and Structural Engineering, Waste and Supplementary Cementitious Materials in Concrete, edited by Rafat Siddique and Paulo Cachim. Woodhead Publishing. 493–511, doi.org/10.1016/B978-0-08-102156-9.00015-8.

34. ASTM C618. 2019. Standard Specification for Coal Fly Ash and Raw or Calcined Natural Pozzolan for Use in Concrete.

35. Bapat, J. 2013. Mineral Admixtures in Cement and Concrete. Boca Raton: CRC Press, 310 p. doi.org/10.1201/b12673.

36. Ptáček, Petr, D. Kubátová, J. Havlica, J. Brandštetr, F. Šoukala and T. Opravila. 2010. "Isothermal kinetic analysis of the thermal decomposition of kaolinite: The thermogravimetric study." Thermochimica Acta 501(1-2): 24–29. https://doi.org/10.1016/j.tca.2009.12.018.

37. De Silva, P.S. and F.P. Glasser. 1992. Pozzolanic activation of metakaolin. Advances in Cement Research 4(16): 167–178.

38. EN 196-5:2011. Methods of testing cement. Pozzolanicity test for pozzolanic cement.

39. Shvarzman, A., K. Kovler, I. Shamban, G.S. Grader and G.E. Shter. 2002. Influence of chemical and phase composition of mineral admixtures on their pozzolanic activity. Advances in Cement Research 14(1): 35–41.

40. Moropoulou, A., A. Bakolas and E. Aggelakopoulou. 2004. Evaluation of pozzolanic activity of natural and artificial pozzolans by thermal analysis. Thermochimica Acta. 420 (1–2): 135–140, doi. org/10.1016/j.tca.2003.11.059.

41. Wala, D. and G. Rosiek. 2003. Minerały ilaste jako dodatek pucolanowy do cementów hydraulicznych [Clay minerals as pozzolanic additive for hydraulic cements]. Cement Wapno Beton 8/70 (1): 27–33. (in Polish).

42. Curcio, F., B.A. DeAngelis and S. Pagliolico. 1998. Metakaolin as a pozzolanic microfiller for high-performance mortars. Cement and Concrete Research 28(6): 803–809. doi.org/10.1016/S0008-8846(98)00045-3.

43. Sabir, B.B., S. Wild and J. Bai. 2001. Metakaolin and calcined clays as pozzolans for concrete: a review. Cement and Concrete Composites 23(6): 441–454. doi.org/10.1016/S0958-9465(00) 00092-5.

44. Asbridge, A.H., G.V. Walters and T.R. Jones. 1994. Ternary blended concretes- OPC/GGBFS/ metakaolin. Denmark: Concrete Across Borders, pp. 547–557.

45. Donatello, S., M. Tyrer and C.R. Cheeseman. 2010. Comparison of test methods to assess pozzolanic activity. Cement and Concrete Composites 32(2): 121–127. doi.org/10.1016/j. cemconcomp.2009.10.008.

46. Ferraz, E., S. Andrejkovičová, W. Hajjaji, A.L. Velosa, A.S. Silva and F. Rocha. 2015. Pozzolanic activity of metakaolins by French standard of the modified Chapelle test: a direct methodology. Acta Geodyn. Geomater 12(3(179)): 289–298.

47. Badogiannis, E., V.G. Papadakis, E. Chaniotakis and S. Tsivilis. 2004. Exploitation of poor Greek kaolins: strength development of metakaolin concrete and evaluation by means of k-value. Cement and Concrete Research 34(6): 1035–1041.

48. Aysha., H., T. Hemalatha, N. Arunachalam, M.A. Ramachandra and N.R. Iyer. 2014. Assessment of Embodied Energy in the Production of Ultra High Performance Concrete (UHPC). International Journal of Students Research in Technology & Management 2(03): 113–120.

49. Bouzoubaâ, N. and M. Lachemi. 2001. Self-compacting concrete incorporating high volumes of class F fly ash: Preliminary results. Cement and Concrete Research 31(3): 413–420. doi.org/10.1016/ S0008-8846(00)00504-4.

50. Jones, M., M. McCarthy and M. Newlands. 2011. Fly Ash Route to Low Embodied CO_2 and Implications for Concrete Construction. In: T. Robl and T. Adams (Eds.). WOCA Proceedings Papers. University Press of Kentucky, 1–14. http://www.flyash.info/2011/018-Jones-2011.pdf.

51. Global Pigments Market 2019 by Manufacturers, Regions, Type and Application, Forecast to 2024. 2019, 135 p. https://www.fiormarkets.com/report/global-metakaolin-market-2019-by-manufacturers-regions-type-366272.html.

52. Thankam, G.L. and N.T. Renganathan. 2020. Ideal supplementary cementing material – Metakaolin: A review. International Review of Applied Sciences and Engineering 11(1): 58–65.

53. Matousek, M. and Z. Šauman. 1974. Contribution to the hydration of expansive cement on the basis of metakaolinite. Cement and Concrete Research 4(1): 113–122. doi.org/10.1016/0008-8846(74)90070-2.

54. Sheikin, A.E. and T.Y. Yakub. 1966. Bezusadochnyi Portland cement [Non shrink cement]. Moscow: Strojizdat, 103 p. (in Russian).

55. Caldarone, M.A., K.A. Gruber and R.G. Burg. 1994. High Reactivity Metakaolin (HRM): A New Generation Mineral Admixture for High Performance Concrete. Concrete International 16(11) 37–41.

56. Wang, J., C. Zhang, J. Xu, P. Qu, Y. Zhou and H.M. Han. 2012. The Effect of Alkali on Compressive of Metakaolin Based Geopolymeric Cement. Advanced Materials Research 554–556 (July 2012): 327–30. doi.org/10.4028/www.scientific.net/amr.554-556.327.

57. Rovnaník, P. 2010. Effect of curing temperature on the development of hard structure of metakaolin-based geopolymer. Construction and Building Materials 24(7): 1176–1183. doi.org/10.1016/j. conbuildmat.2009.12.023.

58. Liew, Y.M., H. Kamarudin, A.M. Mustafa Al Bakri, M. Luqman, I. Khairul Nizar and C.Y. Heah. 2011. Investigating the possibility of utilization of kaolin and the potential of metakaolin to produce

green cement for construction purposes—A review. Australian Journal of Basic and Applied Sciences 5(9): 441–449.

59. Rakhimova, N.R. and R.Z. Rakhimov. 2014. A review on alkali-activated slag cements incorporated with supplementary materials. Journal of Sustainable Cement-Based Materials. 3(1): 61–74. doi: 10.1080/21650373.2013.876944.

60. Krivenko, P., O. Petropavlovskyi and O. Kovalchuk. 2018. A comparative study on the influence of metakaolin and kaolin additives on properties and structure of the alkali-activated slag cement and concrete. Eastern European Journal of Enterprise Technologies 1(6) 33–39. doi.org/10.15587/1729-4061.2018.119624.

61. Kinuthia, J.M., S. Wild, B.B. Sabir and J. Bai. 2000. Self-compensating autogenous shrinkage in Portland cement—metakaolin—fly ash pastes. Advances in Cement Research 12(1): 35–43. doi.org/10.1680/adcr.2000.12.1.35.

62. Frías, M., M.I. Sánchez de Rojas and J. Cabrera. 2000. The effect that the pozzolanic reaction of metakaolin has on the heat evolution in metakaolin-cement mortars. Cement and Concrete Research 30(2): 209–216. doi.org/10.1016/S0008-8846(99)00231-8.

63. Li, Z. and Z. Ding. 2003. Property improvement of Portland cement by incorporating with metakaolin and slag. Cement and Concrete Research 33(4): 579–584. doi.org/10.1016/S0008-8846(02)01025-6.

64. El-Diadamony, H., A.A. Amer, T.M. Sokkary and S. El-Hoseny. 2018. Hydration and characteristics of metakaolin pozzolanic cement pastes. HBRC Journal 14(2): 150–158. doi.org/10.1016/j.hbrcj.2015.05.005.

65. Kuzielová, E., M. Žemlička, E. Bartoničková and M.T. Palou. 2017. The correlation between porosity and mechanical properties of multicomponent systems consisting of Portland cement–slag–silica fume–metakaolin. Construction and Building Materials 135: 306–314. doi.org/10.1016/j.conbuildmat.2016.12.105.

66. Sanytsky, M., T. Kropyvnytska and R. Kotiv. 2014. Modified Plasters for Restoration and Finishing Works. Advanced Materials Research 923: 42–47. doi.org/10.4028/www.scientific.net/amr.923.42.

67. Nwaubani, S., L. Dvorkin and N. Lushnikova. 2011. Influence of metakaolin admixture on mechanical properties and porosity parameters of fiber reinforced concrete. Proceedings of 31st Cement and Concrete Science Conference Novel Developments and Innovation in Cementitious Materials. Imperial College London, 12–13 September: 257–262.

68. Madandoust, R. and S. Yasin Mousavi. 2012. Fresh and hardened properties of self-compacting concrete containing metakaolin. Construction and Building Materials 35: 752–760. doi.org/10.1016/j.conbuildmat.2012.04.109.

69. Dvorkin, L. and N. Lushnikova. 2009. Properties of gypsum binders modified with complex admixtures in Ibausil - Internationale Baustofftagung, Tagunsbericht. Weimar: Institut für Baustoffkunde, Band 2: 1-0701 – 1-0707.

70. Dvorkin, L.Y., O.L. Dvorkin, O.V. Bezusyak, N.V. Lushnikova and I.V. Kovalyk. 2012. Design strategy of foam gypsum proportioning in Ibausil - Internationale Baustofftagung, Tagunsbericht. Weimar: Institut für Baustoffkunde. Band 1.: S. 1-0986 – 1-0992.

71. Vimmrová, A., M. Keppert, O. Michalko and R. Černý. 2014. Calcined gypsum–lime–metakaolin binders: Design of optimal composition. Cement and Concrete Composites 52: 91–96. doi.org/10.1016/j.cemconcomp.2014.05.011.

72. Gameiro, A., A. Santos Silva, P. Faria, J. Grilo, T. Branco, R. Veiga and A. Velosa. 2014. Physical and chemical assessment of lime–metakaolin mortars: Influence of binder:aggregate ratio. Cement and Concrete Composites 45: 264–271. doi.org/10.1016/j.cemconcomp.2013.06.010.

73. Palou, M., E. Kuzielová, M. Žemlička, R. Novotný and J. Másilko. 2018. The effect of metakaolin upon the formation of ettringite in metakaolin–lime–gypsum ternary systems. Journal of Thermal Analysis and Calorimetry 133: 77–86. doi.org/10.1007/s10973-017-6885-0.

74. Fortes-Revilla, C., S. Martínez-Ramírez, M. T. Blanco-Varela. 2006. Modelling of slaked lime–metakaolin mortar engineering characteristics in terms of process variables. Cement and Concrete Composites 28(5): 458–467. doi.org/10.1016/j.cemconcomp.2005.12.006.

75. Zhang, Y., F. Pan and R. Wu. 2016. Study on the performance of FGD gypsum-metakaolin-cement composite cementitious system. Construction and Building Materials 128: 1–11, doi.org/10.1016/j.conbuildmat.2016.09.134.

76. Siddique, R. 2008. Metakaolin. Chapter In: Waste Materials and By-Products in Concrete. Springer, Berlin: Heidelberg, 41–92. doi.org/10.1007/978-3-540-74294-4_2.

77. Lazaar, K., W. Hajjaji, B. Moussi, F. Rocha, J. Labrincha and F. Jamoussi. 2020. Metakaolin and demolition wastes in eco-based sand consolidated concrete. Boletín de la Sociedad Española de Cerámica y Vidrio. April. doi.org/10.1016/j.bsecv.2020.02.004.

78. Zénabou, N.M., N. Gouloure, B. Nait-Ali, S. Zekeng, E. Kamseu, U.C. Melo, D. Smith and C. Leonelli. 2015. Recycled natural wastes in metakaolin based porous geopolymers for insulating applications. Journal of Building Engineering 3: 58–69. doi.org/10.1016/j.jobe.2015.06.006.

79. Melo, C.R., E. Angioletto, H.G. Riella, M. Peterson, M.R. Rocha, A.R. Melo, L. Silva and S. Strugale. 2012. Production of metakaolin from industrial cellulose waste. Journal of Thermal Analysis and Calorimetry 109: 1341–1345. doi.org/10.1007/s10973-011-1892-z.

80. Kapryelov, S.S. and A.V. Sheilfeld. 2017. Nekotorye osobennosti mekhanizma dejstviya organo-mineralnykh modifikatorov na czementnyye sistemy [Certain peculiarties of action mechanism organo-mineral modifiers on cement systems]. Beton i Zhelezobeton 1: 40–46. (in Russian).

81. Kaprielov, S.S. 1995. Obshchye zakonomernosty formyrovanyia struktury tsementnoho kamnia i betona s dobavkoi ultradyspersnykh materyalov [General regularities of structure forming of cement paste and concrete with ultrafine admixtures]. Beton i Zhelezobeton, 1995; 4: 16–20. (In Russian).

82. MetaMax® High quality pozzolanic cement additive. Accessed December 1 2020. https://kaolin.basf.com/products/application/metamax.

83. Fiore, S., F.J. Huertas, F. Huerta and J. Linares. 1995. Morphology of Kaolinite Crystals Synthesized Under Hydrothermal Conditions. Clays and Clay Minerals volume 43: 353–360. doi.org/10.1346/CCMN.1995.0430310.

84. Kenne Diffo, B.B., A. Elimbi, M. Cyr, J. Dika Manga and H. Tchakoute Kouamo. 2015. Effect of the rate of calcination of kaolin on the properties of metakaolin-based geopolymers. Journal of Asian Ceramic Societies 3(1): 130–138. doi.org/10.1016/j.jascer.2014.12.003.

85. Khan, M.I., H.U. Khan, K. Azizli, S. Sufian, Z. Man, A.A. Siyal, N. Muhammad and M.F. ur Rehman. 2017. The pyrolysis kinetics of the conversion of Malaysian kaolin to metakaolin. Applied Clay Science 146: 152–161. doi.org/10.1016/j.clay.2017.05.017.

86. Batrakov, V.G. 1998. Modifitsyrovannye betony. Teoryia i Praktika [Modified concrete. Theory and practice. The 2nd edition]. Moscow: Tekhnoproekt, 768 p. (In Russian).

87. Dvorkin, L.I., O.L. Dvorkin, Y.A. Korneichuk. 1998. Effektyvnye tsementno-zolnye betony [Effective cement-fly ash concrete]. Rovno, 196 p. (In Russian).

88. Sizov, V.P. 1979. Proektirovanie sostavov tyazhelogo betona [Proportioning of heavy-weigth concrete compostions]. Moscow: Strojizdat, 144 p. (In Russian)

89. Okamura, H., K. Ozawa and M. Ouchi. 2000. Self-compacting concrete. Structural Concrete 1(1): 3–17.

90. Gibbs, J., P. Domone, G. De Schutter, P.J.M. Bartos. 2008. Self-Compacting Concrete, 1st ed. Dunbeath, Scotland, UK: Whittles Publishing, 296 p.

91. Nawy, E.G. 1996. Fundamentals of high-strength high-performance concrete. Harlow: Longman Group Limited, 360 p.

92. ASTM C494 / C494M – 19. Standard Specification for Chemical Admixtures for Concrete.

93. Usherov-Marshak, A.V., O.A. Zlatkovskii and M. Ciak. 2004. Assessing the efficiency of chemical and mineral admixtures in early cement hydration. Inorganic Materials [Neorganicheskie Materialy] 40(8): 886–890.

94. Batrakov, V.G., S.S. Kaprielov, F.M. Ivanov and A.V. Sheinfeld. 1990. Oczenka ul'tradispersnykh otkhodov metallurgicheskikh proizvodstv kak dobavok v beton [Estimation of ultrafine metallurgical industry wastes as admixtures to concrete]. Beton i Zhelezobeton 12: 15–17. (in Russian).

95. Cyr, M., P. Lawrence and E. Ringot. 2006. Efficiency of Mineral Admixtures in Mortars: Quantification of the Physical and Chemical Effects of Fine Admixtures in Relation with Compressive Strength. Cement and Concrete Research 36(2): 264–277.

96. Dvorkin, L., A. Bezusyak, N. Lushnikova and Y. Ribakov. 2012. Using mathematical modeling for design of self-compacting high strength concrete with metakaolin admixture. Construction and Building Materials 37: 851–864.

97. Lushnikova, N. 2015. Optimization of selection process of constituent materials for high performance concrete and mortars. Budownictwo i Architektura 14(1): 53–64.

98. ASTM C230. Flow Table for Use in Tests of Hydraulic Cement.

99. Lazniewska-Piekarczyk, B. and J. Gołaszewski. 2019. Relationship Between Air-Content in Fresh Cement Paste, Mortar, Mix and Hardened Concrete Acc. to PN-EN 480-1 with Air-Entraining CEM II/B-V. IOP Conf. Series: Materials Science and Engineering 471(3), 11 p. doi:10.1088/1757-899X/471/3/032044.

100. Dilsa, J., V. Boelb and G. De Schuttera. 2013. Influence of cement type and mixing pressure on air content, rheology and mechanical properties of UHPC. Construction and Building Materials 41: 455–463.

101. Kerkhoff, B. 2002. Benefits of Air Entrainment in HPC. HPC Bridge Views 23: 3.

102. Collepardi, M., S. Collepardi and R. Troli. 2009. Il nuovo calcestruzzo. Tintoretto, 533 p.

103. ASTM C595/C595M-20. Standard Specification for Blended Hydraulic Cements.

104. Akhverdov, I.N. 1981. Osnovy fiziki betona [Concrete Physics Fundamentals]. Moscow: Strojizdat, 464 p. (In Russian).

105. Grushko, I.M., N.F. Glushhenko and A.G. Il`in. 1965. Struktura i prochnost` dorozhnogo czementnogo betona [Structure and strength of road concrete]. Kharkov: Izdatel`stvo Kharkovskogo universiteta, 135 p. (In Russian).

106. Stolnikov, V.V. 1962. Issledovaniya po gidrotekhnicheskomu betonu [Studies on hydrotechnical concrete]. Moscow: Gosenergoizdat, 330 p. (In Russian).

107. Silina, E.S., A.V. Sheinfeld, N.F. Zhigulev and S.T. Borygin. 2000. Svojstva betonnykh smesej s modifikatorom betona MB-01 [Properties of fresh concrete with concrete modifying admixture MB-01]. Beton i Zhelezobeton. 1:3–6. (In Russian).

108. Hewlett, Peter, ed. 2003. Lea's Chemistry of Cement and Concrete. 4th Edition. Imprint: Butterworth-Heinemann, 1092 p.

109. Dzenis, V.V. and V.K. Lapsa. 1971. Ultrazvukovoj kontrol tverdeyushhego betona [Ultrasonic control of hardening concrete]. Leningrad: Strojizdat, 112 p. (In Russian).

110. Wang, B.M., H.N. Ma, M. Li and Y. Han. 2013. Effect of Metakaolin on the Physical Properties and Setting Time of High Performance Concrete. Key Engineering Materials 539: 195–199. doi. org/10.4028/www.scientific.net/kem.539.195.

111. Dvorkin, L.I. and V.P. Kizima. 1986. Effektvnye litye betony. [Effective cast concrete]. Lviv: Vyshcha shkola, 144 p. (In Russian).

112. Sychev, M.M. 1987. Perspektivy povysheniya prochnosti czementnogo kamnya. [Perspectives of strength increasing of hardened cement paste]. Cement 9: 17–19. (In Russian).

113. Rebinder, P.A. and N.A. Cemenenko. 1949. O metode pogruzheniya konusa dlya kharakteristiki strukturno-mekhanicheskikh svojstv plastichno-vyazkikh tel [The method of dipping the cone to characterize structural and mechanical properties of the plastic-viscous bodies]. Doklady Akademii Nauk SSSR [Proceedings of the USSR Academy of Science]. LXIV(6): 835–838. (In Russian).

114. McCarter, W.J. and H.C. Ezirim. 1998. Monitoring the early hydration of pozzolan-Ca(OH)$_2$ mixtures using electrical methods. Advances in Cement Research. 10(4): 161–168.

115. Khigerovich, M.I. and A.P. Merkin. 1968. Fiziko-khimicheskie metody issledovaniya stroitelnykh materialov. [Physical and chemical methods of construction materials researches]. Moscow: Vysshaya shkola, 192 p. (In Russian).

116. Sheikin, A.E., Yu. V. Chekhovskij and M.I. Brusser. 1979. Struktura i svojstva czementnykh betonov. Moscow: Strojizdat, 344 p. (In Russian).

117. Ash, J.E., M.G. Hall, J.I. Langford and M. Mellas. 1993. Estimations of degree of hydration of Portland cement pastes. Cement and Concrete Research 23(2): 399–406. doi.org/10.1016/0008-8846(93)90105-I.

118. Malhotra, V.M. 1989. Superplasticizers: A global review with emphasis on durability and innovative concrete. In. ACI SP-119: Superplasticizers and other chemical admixtures in concrete. Detroit: American Concrete Institute: 1–18.

119. Raoufi, K., M. Pour-Ghaz, A. Poursaee and J. Weiss. 2011. Restrained Shrinkage Cracking in Concrete Elements: Role of Substrate Bond on Crack Development. Journal of Materials in Civil Engineering 23(6): 895–902.

120. Menu, B., B. Bissonnette, M. Jolin and N. Ginouse. 2017. Evaluation of early age shrinkage cracking tendency of concrete. Proceedings of the CSCE Annual Conference on Leadership in Sustainable Infrastructure. Vancouver, Canada. EMM649-1-EMM649-8.

121. Wild, S., J.M. Khatib and L.J. Roose. 1998. Chemical shrinkage and autogenous shrinkage of Portland cement – metakaolin pastes. Advances in Cement Research 10(3): 109–119.
122. Aïtcin, P.-C. and R. Flatt. 2016. Science and Technology of Concrete Admixtures, 1st ed. Woodhead Publishing, 666 p.
123. Lu, D., J. Luo and Z. Xu. 2015. Effect of Metakaolin on the Drying Shrinkage Behaviour of Portland Cement Pastes. In book: Calcined Clays for Sustainable Concrete ed. by K. Scrivener and A. Favier. Dordrecht: Springer 10: 569.
124. Gleize, P., M. Cyr and G. Escadeillas. 2007. Effects of metakaolin on autogenous shrinkage of cement pastes. Cement and Concrete Composites 29(2): 80–87.
125. Jianxia, S. 2012. Durability design of concrete hydropower structures. In book Comprehensive Renewable Energy ed. by T. Letcher. Elsevier, 4422 p.
126. Brooks, J.J., M. Johari and M. Azmi. 2001. Effect of metakaolin on creep and shrinkage. Cement and Concrete Composites 23: 495–502.
127. Güneyisi, E., M. Gesoğlu, K. Mermerdaş. 2008. Improving strength, drying shrinkage and pore structure of concrete using metakaolin. Materials and structures 41(5) pp. 937–949.
128. Poppe, A.-M. and G. De Schutter. 2005. Creep and shrinkage of self-compacting concrete. International Symposium on Design, Performance and Use of Self-Consolidating Concrete, RILEM Proceedings: PRO (42) pp. 329–336.
129. Dvorkin, L.I. and O.L. Dvorkin. 2016. Raschetnoe prognozirovanie svojstv i proektirovanie sostavov betonov. Uchebno-prakticheskoe posobie. [Computational prediction of properties and design of concrete compositions. Educational and practical guide]. Moscow: Infra-Inzheneriya, 384 p. (In Russian).
130. BS EN 206:2013+A1:2016. Concrete. Specification, performance, production and conformity.
131. Mkhitaryan, N.M., G.V. Badeyan and E.G. Malaczidze. 2003. Osobennosti primeneniya tovarnoj betonnoj smesi v monolitnom domostroenii. [Features of the use of commercial concrete in monolithic residential construction]. Stroitelnye materialy i izdeliya. [Construction materials and products] 3: 33–35. (In Russian).
132. Okamura, H. and M. Ouchi. 2003. Self-Compacting Concrete. Journal of Advanced Concrete Technology 1: 5–15.
133. Brusser, M.I., S.A. Vysoczkij and A.M. Czarik. 1989. Opredelenie sokhranyaemosti udoboukladyvaemosti betonnykh smesej. [Determination of retention of workability fresh concrete]. Beton i zhelezobeton [Concrete and Reinforced Concrete] 3: 9–11. (In Russian).
134. Ardalan, R.B., A. Joshaghani and R.D. Hooton. 2017. Workability retention and compressive strength of self-compacting concrete incorporating pumice powder and silica fume. Construction and Building Materials 134: 116–122. doi.org/10.1016/j.conbuildmat.2016.12.090.
135. Runova, R.F., M.O. Kochevyh and I.I. Rudenko. 2005. On the slump loss problem of the superplasticized concrete mixes. Admixtures - Enhancing Concrete Performance. January: 149–156.
136. Punagin, V.N. 1977. Tekhnologiya betona v usloviyakh sukhogo i zharkogo klimata. [Concrete technology in dry and hot climates]. Tashkent: Fan. (In Russian).
137. Ardalan, R.B., A. Joshaghani and R.D. Hooton. 2017. Workability retention and compressive strength of self-compacting concrete incorporating pumice powder and silica fume. Construction and Building Materials 134: 116–122. doi.org/10.1016/j.conbuildmat.2016.12.090.
138. Dvorkin, L., O. Dvorkin and Y. Ribakov. 2012. Mathematical experiments planning in concrete technology. New York: Nova Science Publishers, Inc, 173 p.
139. Montgomery, D.C. 2000. Design and analysis of experiments, 5th ed. New Jersey: Wiley, 688 p.
140. Kolousek, G.L. 1976. Proczessy gidratacii na rannikh stadiyakh tverdeniya czementa [Hydration processes in the early stages of cement hardening]. In book: Sbornik trudov VI mezhdunarodnogo kongressa po khimii czementa. [Proceedings of the VI International Congress on the Chemistry of Cement]. Moscow: Strojizdat. 2(2): 65–79. (In Russian).
141. Box, G.E.P., J.S. Hunter and W.G. Hunter. 2005. Statistics for experimenters: design, discovery and innovation. 2nd ed. Wiley: New Jersey, 672 p.
142. Plank, J., D. Vlad, A. Brandl and P. Chatziagorastou. 2005. Colloidal chemistry examination of the steric effect of polycarboxylate superplasticizers. Cement International 3: 100–10.
143. Ramachandran, V.S. 1995. Concrete admixtures handbook: properties, science and technology. 2nd ed. New Jersey: Noyes Publications, 1183 p.

144. Ismeik, M. 2009. Effect of mineral admixtures on mechanical properties of high strength concrete made with locally available materials. Effect of mineral admixtures on mechanical properties of high strength concrete made with locally available materials. Jordan Journal of Civil Engineering 3(1): 78–90.

145. Akhverdov, I.N. 1961. Vysokoprochnyj beton. [High strength concrete]. M.: Strojizdat, 163 p. (In Russian).

146. Zatvornitskaya, T.A., S.A. Konyaeva and B.F. Mikulovich. 1974. Litye betony v gidroenergeticheskom stroitelstve. [Cast concrete in hydropower construction.] Moscow: Energiya, 112 p. (In Russian).

147. Laidani, Z.E.-A., B. Benabed, R. Abousnina, M.K. Gueddouda and E.-H. Kadri. 2020. Experimental investigation on effects of calcined bentonite on fresh, strength and durability properties of sustainable self-compacting concrete. Construction and Building Materials 230, 117062. doi. org/10.1016/j.conbuildmat.2019.117062.

148. Promsawat, P., B. Chatveera, G. Sua-iam and N. Makul. 2020. Properties of self-compacting concrete prepared with ternary Portland cement-high volume fly ash-calcium carbonate blends. Case Studies in Construction Materials 13: e00426. doi.org/10.1016/j.cscm.2020.e00426.

149. Christodoulou, G. 2000. A comparative study of the effects of silica fume, metakaolin and PFA on the air content of fresh concrete. SCI lecture papers series. London: Society of Chemical Industry: 21–38.

150. Blanks, R.F. and H.L. Kennedy. 1955. The Technology of Cement and Concrete, Vol. 1. New York: Wiley, 422 p.

151. Bazhenov, Yu. M. 1975. Sposoby opredeleniya sostava betona razlichnykh vidov [Methods for determining the composition of concrete of various types]. Moscow: Strojizdat, 272. (In Russian).

152. 211.4R-93: Guide for Selecting Proportions for High-Strength Concrete with Portland Cement and Fly Ash (Reapproved 1998).

153. EN 1992-1-1 Eurocode 2: Design of concrete structures - Part 1-1 : General rules and rules for buildings. Accessed November 25, 2020. https://www.phd.eng.br/wp-content/uploads/2015/12/ en.1992.1.1.2004.pdf.

154. Mikulsky, V.G. and L.A. Igonin. 1965. Sczeplenie i skleivanie betona v sooruzheniyakh. [Adhesion and bonding of concrete in structures]. Moscow: Strojizdat, 128 p. (In Russian).

155. Gordon, S.S. 1966. Struktura betona i jego prochnost s uchetom roli zapolnitelej. [The structure of concrete and its strength taking into account the role of aggregates]. In book: Struktura, prochnost i deformaczii betonov [Structure, strength and deformation of concrete] edited by A.E. Desov. Moscow: Strojizdat. 272–289. (In Russian).

156. Kjellsen, K.O., O.H. Wallevik and L. Fjällberg. 1998. Microstructure and microchemistry of the paste-aggregate interfacial transition zone of high-performance concrete. Advances in Cement Research 10(1): 33–40.

157. Bentz, D.P. and E.J. Garboczi. 1991. Simulation studies of the effects of mineral admixtures on the cement paste – aggregate interfacial zone. ACI Materials Journal 8: 518–529.

158. Menon, A., C.M. Childs, B. Poczós, N.R. Washburn and K.E. Kurtis. 2019. Molecular Engineering of Superplasticizers for Metakaolin–Portland Cement Blends with Hierarchical Machine Learning. Advanced Theory and Simulations 2(4).

159. Sathyan, D., K.B. Anand, K.M. Mini and S. Aparna. 2017. Optimization of superplasticizer in portland pozzolana cement mortar and concrete. Published under licence by IOP Publishing Ltd IOP Conference Series: Materials Science and Engineering, International Conference on Advances in Materials and Manufacturing Applications (IConAMMA-2017) 17–19 August, Vol. 310, Bengaluru, India.

160. Bentz, D.P. and P.E. Stutzman. 2006. Curing, hydration and microstructure of cement paste. ACI Materials Journal Sep/Oct: 348–356.

161. Dinh Dau, V. and P. Stroeven. 2003. Strength improvement efficiency of mineral admixtures in concrete. Proceedings of the International Conference on Advances in Concrete and Structures. RILEM Publications SARL: 785–792.

162. Ganesh Babu, K. and P.V. Surya Prakash. 1995. Efficiency of silica fume in concrete. Cement and Concrete Research 25(6): 1273–1283.

163. Papadakis, V.G. and S. Tsimas. 2002. Supplementary cementing materials in concrete Part I: efficiency and design. Cement and Concrete Research 32: 1525–1532.

164. Boudchicha, A., M.C. Zouaoui, J.L. Gallias and B. Mezghiche. 2007. Analysis of the effects of mineral admixtures on the strength of mortars: Application of the predictive model of Feret. Journal of Civil Engineering and Management 13(2): 87–96.

165. Wong, H.S. and H.A. 2005. Razak Efficiency of calcined kaolin and silica fume as cement replacement for strength performance. Cement and Concrete Research 35(4): 696–702.

166. Dinakar, P. 2011. High reactive metakaolin for high strength and high performance concrete. Indian Concrete Journal April: 28–34.

167. Bharatkumar, B.H., R. Narayanan, B.K. Raghuprasad and D.S. Ramachandramurthy. 2001. Mix proportioning of high performance concrete. Cement and Concrete Composites 23(1): 71–80.

168. Iskhakov, I., Y. Ribakov and A.A. Shah. 2009. Experimental and Theoretical Investigation of Column - Flat Slab Joint Ductility. Materials and Design 30: 3158–3164.

169. Farzadnia, N., A.A.A. Ali and R. Demirboga. 2011. Incorporation of Mineral Admixtures in Sustainable High Performance Concrete. International Journal of Sustainable Construction Engineering & Technology 2(1): 44–56.

170. Thilagavathi, S., G. Dhinakaran and J. Venkata Ramana. 2007. Effects of Mineral Admixtures, Water Binder Ratio and Curing on Compressive Strength of Concrete. Journal of Civil Engineering Research and Practice 4 (2): 31–42.

171. Berg, O. Ya., E.N. Shherbakov and G.N. Pisanko. 1971. Vysokoprochnyj beton. [High strength concrete]. Moscow: Strojizdat, 208 p. (In Russian)

172. Walter, H.D. and W. Changqing. 1995. Shrinkage and Creep of High-Performance Concrete (HPC). A Critical Review. Proceedings of Symposium on Concrete Technology. Las Vegas, 212–216.

173. Issers, F.A., M.G. Bulgakova and N.I. Vershinina. 1999. Prochnostnye i deformativnye svojstva vysokoprochnykh betonov s modifikatorom MB-01. [Strength and deformation properties of high-strength concretes with modifier MB-01]. Beton i zhelezobeton [Concrete and Reinforced Concrete] 3: 6–9. (In Russian).

174. Dorofeev, V.S. and V.N. Vyrovoj. 1998. Tekhnologicheskaya povrezhdennost stroitelnykh materialov i konstrukczij [Technological damage to building materials and structures]. Odessa: Gorod masterov. 165 p. (In Russian).

175. Arioglu, N., Z.C. Girgin and E. Arioglu. 2006. Evaluation of ratio between splitting tensile strength and compressive strength for concretes up to 120 MPa and its application in strength criterion. ACI Materials Journal 103(1):18–24.

176. Rushhuk, G.M. 1957. O svyazi mezhdu prochnostnymi kharakteristikami i dinamicheskim modulem uprugosti czementnykh rastvorov. [On the relationship between the strength characteristics and the dynamic modulus of elasticity of cement mortars]. Cement. 2: 24–29. (In Russian).

177. Desov, A.E. 1966. Nekotorye voprosy struktury, prochnosti i deformaczii betonov. [Some questions of structure, strength and deformation of concretes] In book: Struktura, prochnost i deformaczii betonov [Structure, strength and deformation of concrete] edited by A.E. Desov. Moscow: Strojizdat. 4–59. (In Russian).

178. Dvorkin, O.L. 2003. Proektirovanie sostavov betona. Osnovy teorii i metodologii. [Design of concrete compositions. Fundamentals of theory and methodology]. Rovno: UDUWGP, 266 p. (In Russian).

179. Krivenko, P.V. 2002. Sovremennye problemy dolgovechnosti betona: sostoyanie i perspektivy. [Modern problems of concrete durability: state of art and perspectives]. In book Suchasni problemi betonu ta jogo tekhnologij [Proceedings Current problems of concrete and its technologies"]. Kyiv: NDIBK. 56: 15–18. (In Russian).

180. Halbiniak, Jacek and Bogdan Langier. 2014. The Characterization of Porosity and Frost Resistance of Concrete with Fly Ashes Modified. Advanced Materials Research 1020: 193–98. doi.org/10.4028/www.scientific.net/amr.1020.193.

181. Dehn, F. 2000. Heller hochfester Beton unter Verwendung von Metakaolin [Light-colored high-strength concrete using metakaolin]. Lacer. 5: 141–146. (in German)

182. Pavlenko, A., A. Mishakova, O. Pertseva, V. Ivanova, Y. Olekhnovich and G. Averchenko. 2020. Feasibility of using of accelerated test methods for determination of frost-resistance for concrete. E3S Web Conf., 157. doi: https://doi.org/10.1051/e3sconf/202015706035.

183. ASTM C666 / C666M – 15. Standard Test Method for Resistance of Concrete to Rapid Freezing and Thawing.

184. Gorchakov, G.I., L.P. Orentlikher and L.I. Savin. 1976. Sostav, struktura i svojstva czementnykh betonov. [Composition, structure and properties of cement concrete]. Moscow: Strojizdat, 145 p. (In Russian).

185. Sizov, V.P. 1992. Prognozirovanie morozostojkosti betona. [Forecasting frost resistance of concrete]. Beton i zhelezobeton [Concrete and Reinforced Concrete] 6: 25–27. (In Russian).

186. Sizov, V.P. 1996. K voprosu prognozirovaniya vodoneproniczaemosti betona. [On the issue of forecast watertightness of concrete]. Beton i zhelezobeton [Concrete and Reinforced Concrete] 6:23–24. (In Russian).

187. Koval, S.V. 2003. Povyshenie effektivnosti v ispolzovanii dobavok v tekhnologii betona na osnove modelirovaniya i kompyuternogo poiska optimalnykh reczeptur. [Increasing the effectiveness in application of admixtures in technology of concrete based on design and computerized selection of optimal proportions]. Stroitelnye materialy i izdeliya [Construction materials and products] 6: 26–28. (In Russian).

188. ACI Committee 363. 2005. High-Strength Concrete (ACI 363R).228: 79-80. 6/1/2005.

189. Kurdovski, W and H. Pomadowski. 2001. Influence of Portland cement composition on pozzolanic reactivity of metakaolin. Silicates Industriels 66 (7-8): 85–90.

190. Ta Min` Khoang. 2001. Melkozernistyj beton s dobavkoj metakaolina [Fine-grained concrete with metakaolin admixture]. Stroitelnye materialy, oborudovanie, tekhnologii XXI veka. [Construction materials, equipment, technology of XXI century] 11: 13.

191. Usherov-Marshak, A., V. Sopov, S. Kaprielov and H. Kardumian. 2006. Influence of organic-mineral admixtures on early hydration of cements". Ibausil - Internationale Baustofftagung, Tagungsbericht. Weimar: Institut für Baustoffkunde. Band 1: 1-0653 – 2-0660.

192. Ratynov, V.B. and T.Y. Rozenberh. 1989. Dobavky v beton [Concrete admixtures]. Moscow: Strojizdat, 188 p. (In Russian).

193. Mal, J. and J. Dietzl. 2002. Ultra High Performance Self Compacting Concrete. Laser, 7: 33–42.

194. ACI 211.1-91. 1991. Standard Practice for selecting proportions of normal, heavy-weight and mass concrete. ACI Committee 211 Report, American Concrete Institute.

195. Gass, S.I. 1975. Linear programming. 3rd ed. New York: McGraw-Hill Inc, 532 p.

196. Halstead, P.E. and S.D. Lawrence. 1964. Kinetika Reaktsiy v Sisteme CaO·SiO₂·N₂O. [Kinetics of Reactions in the CaO·SiO₂·H₂O System]. Proceedings of the 4th Congress on the Cement Chemistry. Moscow: Stroyizdat. p. 261–264. (In Russian).

197. Uchikawa, H. 1986. Effect of Blending Components of Hydration and Structure Formation. 8-th International Congress on the Chemistry of Cement. Vol.1. Rio de Janeiro. p. 250–280.

198. Richard, P. and M. Cheyrezy. 1995. Composition of reactive powder concretes. Cement and Concrete Research 25(7): 1501–1511.

199. Long, G.C., X.Y. Wang and Y.J. Xie. 2002. Very-high-performance concrete with ultrafine powders. Cement and Concrete Research 32: 601–605.

200. Morin, V., F. Choen-Tenoudji, A. Feylessoufi and P. Richard. 2002. Evolution of the capillary network in a reactive powder concrete during hydration process. Cement and Concrete Research, 32 (12): 1907–1914.

201. Shaheen, E. and N.G. Shrive. 2006. Optimization of Mechanical Properties and Durability of Reactive Powder Concrete. ACI Materials Journal 103 (6): 444–451.

Index